Neuroscience Step by Step

A Structured Introduction to the Science of the Brain

Frank Connors

© 2024 by Frank Connors

All rights reserved.

No part of this publication may be reproduced, distributed, or transmitted in any form or by any means, including photocopying, recording, or other electronic or mechanical methods, without the prior written permission of the publisher, except for brief quotations in critical reviews and some other noncommercial uses permitted by copyright law.

This book is designed to provide general information regarding the topics discussed. It is offered with the understanding that neither the author nor the publisher is engaged in rendering financial, legal, or other professional advice. While efforts have been made to ensure the accuracy and reliability of the information contained in this publication, the author and publisher do not guarantee its accuracy or completeness and shall not be responsible for any errors, omissions, or for the results obtained from the use of such information. The material in this book is provided "as is," without any express or implied warranties.

Readers are encouraged to consult a qualified professional for advice tailored to their personal or professional situation. The strategies and information discussed may not be appropriate for every situation and are not promised or guaranteed to produce specific outcomes. Neither the author nor the publisher will be liable for any loss, damage, or other consequences that may arise from the use of or reliance on the information provided.

No representation is made about the quality of information provided exceeding that obtainable through professional advice. In no event will the author or publisher be responsible for any direct, indirect, incidental, consequential, or other damages resulting from the use of the information in this book.

This book is not intended as a substitute for professional economic, investment, or other expert advice. It should not be used as the sole basis for any financial decisions or actions that might affect your personal or professional circumstances.

Consultation with a professional advisor is recommended for advice specific to your situation. Any reliance you place on the information in this book is strictly at your own risk.

PREFACE

Welcome, curious reader! You've just opened the door to an incredible journey through the most complex and fascinating organ in the known universe – the human brain. Whether you're a student, a professional, or simply someone intrigued by how our minds work, this book is designed to be your friendly guide through the exciting world of neuroscience.

How the brain work can be an intimidating subject. With its billions of neurons, intricate networks, and the ability to generate our thoughts, emotions, and experiences, it's no wonder that understanding the brain can feel like a daunting task. But don't worry! We've created this book to break down the complexities of neuroscience into bite-sized, digestible pieces that anyone can understand and enjoy.

"Neuroscience Step by Step" is exactly what it says on the cover – a structured introduction to the science of the brain. We'll start with the basics and gradually build up to more complex concepts, ensuring that you have a solid foundation before moving on to more advanced topics. At the same time, for those who want to skip around or pick and choose what to read, each chapter and section is designed to stand on its own as well.

We'll go into the microscopic world of neurons and discover how these tiny cells communicate to create the symphony of our conscious experience. You'll learn about the chemicals that allow our brain cells to "talk" to each other and how this communication can go awry in various disorders.

We'll also zoom out to look at the bigger picture, exploring how different parts of the brain work together to process sensory information, control our movements, store memories, and regulate our emotions. You'll gain insights into the mysteries of sleep, the complexities of pain, and the relationship between our nervous and immune systems.

But this book isn't just about understanding how a healthy brain functions. We'll also go into what happens when things go wrong, exploring various neurological and psychiatric disorders. More importantly, we'll discuss the cutting-edge treatments and therapies that neuroscientists and medical professionals are developing to help those affected by these conditions.

One of the most exciting aspects of neuroscience is how rapidly the field is advancing. New discoveries are being made almost daily, pushing the boundaries of our understanding and opening up new possibilities for treating brain disorders. While we can't predict the future, we'll give you a glimpse of where neuroscience might be heading and the potential impacts on medicine, technology, and our understanding of what it means to be human.

Neuroscience is a vast and complex field, and it's natural to need time to absorb and reflect on the information. Feel free to reread sections, jot down questions, or discuss concepts with friends or fellow students. Learning about the brain is a journey, not a race.

We've included an appendix with terms and definitions to help you along the way. Don't hesitate to refer to it if you encounter unfamiliar terms or concepts.

Above all, we hope this book ignites your curiosity and leaves you with a sense of wonder about the incredible organ that makes you, well, you. The human brain is a marvel of nature, and understanding even a fraction of how it works can be a profoundly enriching experience.

TOPICAL OUTLINE

Chapter 1: Foundations of Neuroscience
1. The Evolution of Neuroscience
2. Basic Anatomy of the Brain
3. Neurotransmission Basics
4. Methods in Neuroscience Research
5. Neuroplasticity: The Brain's Ability to Change
6. Neurogenesis: Birth of New Neurons
7. The Role of Genetics in Neuroscience
8. Ethics in Neuroscience

Chapter 2: The Structure and Function of Neurons
1. Neuronal Anatomy
2. Types of Neurons
3. Glial Cells
4. Neuronal Communication
5. Synaptic Structure and Function
6. Neuronal Networks and Circuits

Chapter 3: Neurotransmitters and Their Receptors
1. Major Neurotransmitters
2. Receptor Types
3. Neurotransmitter Systems and Behavior
4. Pharmacology of Neurotransmission
5. Neuromodulators and Their Effects
6. Neurotransmitter Synthesis, Degradation, Transport, and Reuptake

Chapter 4: The Development of the Nervous System
1. Embryonic Development
2. Cell Migration and Axon Guidance
3. Synaptogenesis
4. Critical Periods in Development
5. Developmental Disorders of the Nervous System

Chapter 5: Sensory Systems: How We Perceive the World
1. The Visual System
2. The Auditory System
3. Somatosensory System
4. Olfactory and Gustatory Systems
5. The Vestibular System
6. Proprioception and Kinesthesia

Chapter 6: Motor Systems: From Thought to Action
1. Motor Cortex and Control of Movement
2. Basal Ganglia and Movement Regulation
3. Cerebellum and Coordination
4. Spinal Cord and Reflexes

Chapter 7: Learning and Memory: How We Store Information
1. Types of Memory
2. Neural Circuits of Memory
3. Mechanisms of Synaptic Plasticity
4. Disorders of Memory
5. Consolidation and Reconsolidation
6. Working Memory

Chapter 8: Emotion and the Brain
1. The Limbic System
2. Emotion Regulation and Empathy
3. The Neurochemistry of Emotion
4. Disorders of Emotion

Chapter 9: Cognition and Executive Function
1. The Prefrontal Cortex
2. Attention and Perception
3. Language and the Brain
4. Cognitive Aging
5. Creativity and the Brain
6. Metacognition

Chapter 10: Sleep and Circadian Rhythms
1. Stages of Sleep
2. Neural Mechanisms of Sleep Regulation
3. Circadian Rhythms
4. Sleep Disorders
5. Sleep Across the Lifespan
6. Dreaming and Brain Activity

Chapter 11: Neuroscience of Pain
1. Types of Pain
2. Pain Pathways in the Brain
3. Modulation of Pain
4. Pain Management

Chapter 12: Neuroimmunology: The Brain and the Immune System

1. The Blood-Brain Barrier
2. Neuroinflammation
3. The Gut-Brain Axis
4. Autoimmune Diseases of the Brain
5. Neuroendocrine-Immune Interactions

Chapter 13: Neurodegenerative Diseases

1. Alzheimer's Disease
2. Parkinson's Disease
3. Huntington's Disease
4. Amyotrophic Lateral Sclerosis (ALS)
5. Multiple Sclerosis
6. Prion Diseases
7. Neuroprotective Strategies

Chapter 14: Neuropsychiatric Disorders

1. Schizophrenia
2. Mood Disorders
3. Anxiety Disorders
4. Autism Spectrum Disorder
5. Personality Disorders
6. Addiction and the Brain

Chapter 15: Therapeutic Approaches in Neuroscience

1. Pharmacotherapy
2. Neurosurgery
3. Psychotherapy and Behavioral Interventions
4. Neuromodulation Techniques
5. Cognitive Rehabilitation

Appendix

- Terms and Definitions

Afterword

TABLE OF CONTENTS

Chapter 1: Foundations of Neuroscience ... 1
Chapter 2: The Structure and Function of Neurons ... 17
Chapter 3: Neurotransmitters and Their Receptors ... 25
Chapter 4: The Development of the Nervous System ... 33
Chapter 5: Sensory Systems: How We Perceive the World ... 40
Chapter 6: Motor Systems: From Thought to Action ... 46
Chapter 7: Learning and Memory: How We Store Information ... 51
Chapter 8: Emotion and the Brain ... 58
Chapter 9: Cognition and Executive Function ... 64
Chapter 10: Sleep and Circadian Rhythms ... 72
Chapter 11: Neuroscience of Pain ... 80
Chapter 12: Neuroimmunology: The Brain and the Immune System ... 86
Chapter 13: Neurodegenerative Diseases ... 93
Chapter 14: Neuropsychiatric Disorders ... 102
Chapter 15: Therapeutic Approaches in Neuroscience ... 111
Appendix ... 118
Afterword ... 122

CHAPTER 1: FOUNDATIONS OF NEUROSCIENCE

The Evolution of Neuroscience

Neuroscience, as a field, has a deep history rooted in ancient curiosity about the brain and its functions. The journey began over two millennia ago when early civilizations started questioning the essence of human thought and behavior. The Egyptians, for example, had some knowledge of the brain, evident from their early medical texts. They noted that injuries to the brain could affect behavior, but they largely misunderstood its importance, focusing more on the heart as the seat of the soul.

Hippocrates, the Greek physician from the 5th century BCE, made a significant leap in understanding. He argued that the brain was the source of intelligence and sensation, challenging the prevailing belief that the heart held these functions. This idea laid the groundwork for future explorations into brain function, even though it would take centuries for these ideas to develop further.

The Renaissance period was crucial for the evolution of neuroscience. The invention of the printing press allowed for the dissemination of ideas, and the revival of human dissection provided more detailed knowledge of brain anatomy. Andreas Vesalius, a 16th-century anatomist, produced detailed drawings of the brain, showing structures like the ventricles, which were thought to play a role in the body's fluids and thus in sensation and movement.

However, the true foundation of modern neuroscience didn't emerge until the late 19th century. During this period, scientists began to view the brain not just as a single organ but as a complex system composed of various interconnected parts. **This era marked the shift from viewing the brain as a homogenous mass to understanding it as a network of specialized regions**. Santiago Ramón y Cajal, a Spanish pathologist, was a pioneer in this shift. Using Golgi's silver staining technique, Cajal was able to visualize individual neurons, revealing the intricate network of connections that make up the brain's structure. His work established the **neuron doctrine**, which proposed that neurons are the basic structural and functional units of the brain.

Around the same time, Charles Sherrington introduced the concept of synapses, the junctions between neurons where communication occurs. His work on reflexes demonstrated how neurons work together to produce coordinated responses to stimuli, fundamentally shaping our understanding of how the brain controls behavior. These discoveries underscored the idea that the brain's functions could be understood by studying its cellular and molecular components.

As the 20th century progressed, neuroscience continued to evolve with the development of new technologies and methodologies. The invention of the electron microscope in the 1950s allowed scientists to examine the fine details of neuronal structures at a much higher resolution than ever before. This advancement deepened our understanding of the synapse and how neurotransmitters, the chemicals that facilitate communication between neurons, operate.

Neuroscience's evolution was also influenced by psychology and computer science, particularly in the mid-20th century. The development of computational models of the brain brought a new perspective to understanding neural processes. Researchers began to compare the brain to a computer, leading to the birth of cognitive neuroscience. This interdisciplinary field aims to understand how brain functions produce mental processes like perception, memory, and decision-making.

By the latter half of the 20th century, the advent of neuroimaging techniques like EEG, PET, and fMRI revolutionized neuroscience. These tools allowed scientists to study the brain in action, observing which areas of the brain are active during different tasks. **This ability to observe brain activity in real-time transformed neuroscience from a largely observational science into one that could test hypotheses about how specific brain regions contribute to behavior.**

In recent decades, neuroscience has expanded its scope to include genetics, molecular biology, and more advanced imaging techniques. The discovery of neuroplasticity, the brain's ability to reorganize itself by forming new neural connections throughout life, challenged long-held beliefs that the brain's structure is relatively fixed after childhood. **This discovery has profound implications for understanding recovery from brain injury, the development of mental health disorders, and the aging process.**

Today's neuroscience is a vast, interdisciplinary field that spans everything from molecular studies of ion channels in neurons to the development of artificial neural networks in machine learning. The evolution of neuroscience is ongoing, driven by continuous advancements in technology, cross-disciplinary collaboration, and a deeper understanding of the brain's complexities.

Basic Anatomy of the Brain

The brain, one of the most complex organs in the human body, is the control center for everything we do—whether we're thinking, moving, or feeling. Understanding the basic anatomy of the brain is essential for grasping how it functions. The brain is divided into several major parts, each with distinct structures and roles.

The Cerebrum is the largest part of the brain, accounting for about 85% of its total weight. It's divided into two hemispheres: the left and the right. These hemispheres are connected by a bundle of nerve fibers called the corpus callosum, which allows them to communicate with each other. Each hemisphere controls the opposite side of the body; for example, the left hemisphere controls the right hand. The surface of the cerebrum is called the cerebral cortex, which is made up of folds called gyri and grooves called sulci. These folds increase the surface area, allowing for more neurons and thus higher processing power.

The cerebral cortex itself is divided into four **lobes**, each responsible for different functions:

1. **Frontal Lobe**: Located at the front of the brain, it's involved in decision-making, problem-solving, emotional control, and voluntary movement. The frontal lobe also contains the motor cortex, which controls movement.
2. **Parietal Lobe**: Positioned behind the frontal lobe, it processes sensory information such as touch, temperature, and pain. It also plays a role in spatial orientation and movement coordination.
3. **Occipital Lobe**: Located at the back of the brain, this lobe is primarily responsible for vision. It processes visual information and helps us understand what we see.
4. **Temporal Lobe**: Found on the sides of the brain, it's involved in processing auditory information and is key to understanding language. The temporal lobe also plays a significant role in memory and emotion.

The Cerebellum, situated under the cerebrum at the back of the brain, is much smaller but incredibly important. It coordinates voluntary movements such as posture, balance, and coordination, ensuring that actions are smooth and precise. It also plays a role in motor learning, helping the body to refine movements through practice.

The Brainstem is located beneath the cerebrum and in front of the cerebellum, connecting the brain to the spinal cord. It's composed of three parts: the midbrain, the pons, and the medulla oblongata. The brainstem controls many of the body's basic life functions, such as heart rate, breathing, and blood pressure. It also regulates the sleep-wake cycle and is involved in reflex actions like swallowing and coughing.

1. **Midbrain**: The uppermost part of the brainstem, the midbrain is involved in functions such as vision, hearing, eye movement, and body movement.
2. **Pons**: Situated below the midbrain, the pons acts as a bridge between different parts of the nervous system. It's involved in breathing regulation, facial expressions, and transmitting signals to the cerebellum.
3. **Medulla Oblongata**: The lowest part of the brainstem, it's crucial for regulating vital functions like heart rate, blood pressure, and respiration. It also controls reflexes like swallowing and vomiting.

The Limbic System is a group of interconnected structures located deep within the brain, beneath the cerebral cortex. It's primarily responsible for emotions, memories, and arousal (or stimulation). The key components of the limbic system include:

- **Hippocampus**: Essential for forming new memories and connecting emotions to those memories.
- **Amygdala**: Plays a central role in processing emotions, particularly fear and pleasure.
- **Thalamus**: Acts as a relay station, sending sensory and motor signals to the cerebral cortex. It also plays a role in consciousness and sleep.
- **Hypothalamus**: Located below the thalamus, it regulates body temperature, hunger, thirst, and other autonomic functions. It also controls the pituitary gland, linking the nervous system to the endocrine system.

The Ventricles and Cerebrospinal Fluid (CSF): The brain contains four interconnected cavities called ventricles, which produce and circulate cerebrospinal fluid. This fluid cushions the brain, removes waste, and provides a stable environment for the brain.

Neurotransmission Basics

Neurotransmission is the process by which neurons communicate with each other and with other cells in the body. This communication is essential for every action, thought, and feeling. Understanding the basics of neurotransmission gives insight into how the brain processes information and coordinates responses.

At the heart of neurotransmission are **neurons**, the specialized cells that transmit signals throughout the nervous system. Neurons are composed of three main parts: the cell body (soma), dendrites, and the axon. The cell body contains the nucleus and other organelles, while the dendrites receive incoming signals from other neurons. The axon is a long, thin extension that transmits signals away from the neuron to other cells.

Neurotransmission begins with an **electrical signal** called an action potential. When a neuron is activated by a stimulus (such as a sensory input or a signal from another neuron), an action potential is generated at the axon hillock, a region near the start of the axon. This action potential is an all-or-nothing event, meaning it either happens fully or not at all. Once initiated, the action potential travels rapidly down the axon toward the axon terminal, the end of the neuron.

As the action potential reaches the axon terminal, it triggers the release of **neurotransmitters**, which are chemical messengers. Neurotransmitters are stored in small sacs called vesicles within the axon terminal. When the action potential arrives, it causes these vesicles to fuse with the cell membrane at the synapse, a tiny gap between the axon terminal of one neuron and the dendrite of the next neuron.

Neurotransmitters are then released into the synapse, where they cross the gap and bind to specific receptors on the surface of the neighboring neuron's dendrites. Each neurotransmitter has a specific shape that fits into its corresponding receptor, similar to a key fitting into a lock. This binding process triggers a response in the receiving neuron. Depending on the type of neurotransmitter and receptor involved, the response can either be excitatory (increasing the likelihood that the next neuron will fire an action potential) or inhibitory (decreasing that likelihood).

Some of the most well-known neurotransmitters include:

1. **Glutamate**: The most common excitatory neurotransmitter in the brain, involved in nearly all aspects of brain function, including learning and memory.
2. **GABA (Gamma-Aminobutyric Acid)**: The primary inhibitory neurotransmitter in the brain, crucial for reducing neuronal excitability and preventing overstimulation.
3. **Dopamine**: Associated with reward, motivation, and motor control, dopamine plays a key role in behaviors such as pleasure-seeking and movement.
4. **Serotonin**: Involved in regulating mood, appetite, and sleep. Imbalances in serotonin levels are linked to conditions like depression and anxiety.
5. **Acetylcholine**: Important for muscle contraction, as well as learning and memory. It's the primary neurotransmitter involved in the communication between motor neurons and muscles.
6. **Norepinephrine**: Functions as both a neurotransmitter and a hormone, playing a role in attention, arousal, and the body's "fight or flight" response.

After the neurotransmitter has bound to its receptor and triggered a response, it must be cleared from the synapse to allow the neuron to reset and prepare for the next signal. This clearance can happen in several ways:

- **Reuptake**: The neurotransmitter is taken back into the axon terminal that released it, where it can be repackaged into vesicles for future use.
- **Enzymatic Breakdown**: Enzymes in the synapse break down the neurotransmitter into inactive components.
- **Diffusion**: The neurotransmitter diffuses away from the synapse and is absorbed by surrounding cells.

This entire process of neurotransmission, from the generation of the action potential to the clearance of the neurotransmitter, happens incredibly quickly—often within milliseconds. **It's this rapid and precise communication between neurons that enables the brain to perform complex tasks, such as thinking, moving, and responding to the environment.**

Neurotransmission is also the target of many medications and drugs. For instance, antidepressants like SSRIs (Selective Serotonin Reuptake Inhibitors) work by blocking the reuptake of serotonin, increasing its availability in the synapse and thereby enhancing mood regulation.

Understanding the basics of neurotransmission provides a foundation for exploring more complex brain functions, as well as the mechanisms underlying various neurological and psychiatric disorders.

Methods in Neuroscience Research

Neuroscience research employs a variety of methods to explore the brain's structure, function, and activity. Each method provides unique insights, helping scientists piece together the complex puzzle of how the brain works. Here's a look at some of the key methods used in neuroscience research:

1. Neuroimaging Techniques

Neuroimaging allows researchers to visualize the brain's structure and activity in living organisms. There are several types of neuroimaging methods, each with its own strengths:

- **Magnetic Resonance Imaging (MRI):** MRI uses strong magnetic fields and radio waves to produce detailed images of the brain's anatomy. It's particularly useful for identifying structural abnormalities, such as tumors or lesions.
- **Functional MRI (fMRI):** fMRI measures brain activity by detecting changes in blood flow. When a brain area is more active, it requires more oxygen, which is delivered by increased blood flow. fMRI allows researchers to see which parts of the brain are involved in specific tasks, like memory or decision-making.
- **Positron Emission Tomography (PET):** PET involves injecting a small amount of radioactive material into the bloodstream. This tracer emits positrons, which collide with electrons in the body to produce gamma rays. PET scans detect these gamma rays, allowing researchers to observe metabolic processes in the brain, such as glucose consumption.
- **Electroencephalography (EEG):** EEG measures electrical activity in the brain using electrodes placed on the scalp. It provides a continuous record of brainwave patterns, making it useful for studying processes like sleep, epilepsy, and cognitive functions.

2. Electrophysiology

Electrophysiology involves measuring the electrical activity of neurons. This method can be applied at various levels, from single neurons to large networks of neurons:

- **Single-Unit Recording:** This technique involves inserting a microelectrode into the brain to record the activity of individual neurons. It's commonly used in animal studies to understand how specific neurons respond to stimuli.
- **Patch-Clamp Recording:** This technique allows researchers to study the ion channels on the membrane of a neuron by measuring the currents that flow through them. It provides detailed information about how neurons generate action potentials and transmit signals.
- **Intracranial EEG (iEEG):** Unlike standard EEG, iEEG involves placing electrodes directly on the brain's surface or within the brain tissue. It's used in clinical settings, particularly for identifying the source of seizures in patients with epilepsy.

3. Molecular and Genetic Techniques

These methods are used to study the brain at the molecular and genetic levels, helping researchers understand the biochemical processes underlying neural function:

- **Gene Knockout/Knock-In:** This technique involves creating organisms in which specific genes are either removed (knockout) or inserted/modified (knock-in). By observing the effects of these genetic changes, researchers can learn about the role of specific genes in brain development and function.
- **CRISPR-Cas9:** This revolutionary gene-editing tool allows precise modifications to DNA. In neuroscience, it's used to study the effects of specific genes on brain function, and it holds potential for developing treatments for genetic brain disorders.
- **Optogenetics:** This method involves genetically modifying neurons to express light-sensitive ion channels. By shining light on these neurons, researchers can control their activity with high precision, allowing them to study the function of specific neural circuits in real-time.

4. Behavioral Neuroscience Methods

Behavioral neuroscience focuses on understanding how the brain influences behavior. This often involves a combination of behavioral testing and physiological measurements:

- **Conditioning and Learning Paradigms:** These experiments test how animals or humans learn and remember information. Common methods include classical conditioning, operant conditioning, and maze learning.

- **Lesion Studies:** In animals, specific brain regions can be surgically removed or chemically inactivated to study their role in behavior. In humans, researchers often study individuals with naturally occurring brain lesions to understand the function of the damaged areas.
- **Pharmacological Manipulation:** Drugs that affect neurotransmitter systems can be administered to study their effects on behavior and brain function. This approach helps researchers understand the role of different neurotransmitters in processes like mood regulation and cognition.

5. Computational Neuroscience

Computational neuroscience uses mathematical models and simulations to understand brain function. This method bridges experimental neuroscience and theoretical biology, providing insights that are difficult to obtain through empirical methods alone:

- **Neural Network Models:** These models simulate how networks of neurons interact to produce complex behaviors, such as learning, memory, and decision-making. They are often inspired by real neural circuits but simplified for computational analysis.
- **Data Analysis Algorithms:** With the increasing volume of data from brain research, computational methods are essential for analyzing and interpreting this information. Machine learning algorithms, for example, can help identify patterns in neural activity that correlate with specific behaviors or cognitive states.

6. Neuropsychology

Neuropsychology studies the relationship between brain function and behavior, often through the assessment of individuals with brain injuries or neurological disorders:

- **Cognitive Testing:** Neuropsychologists use standardized tests to measure various aspects of cognition, such as memory, attention, and executive function. These tests help identify areas of the brain that may be impaired and how these impairments affect behavior.
- **Case Studies:** In-depth studies of individuals with specific brain injuries or disorders provide insights into the role of different brain regions. Famous case studies, like that of Phineas Gage, have significantly advanced our understanding of brain function.

These methods in neuroscience research are often used in combination to provide a more comprehensive understanding of how the brain works. Each method contributes a different piece of the puzzle, whether it's mapping brain structures, measuring electrical activity, manipulating genes, or observing behavior. This multi-faceted approach is essential for unraveling the complexities of the brain and developing effective treatments for neurological and psychiatric conditions.

Neuroplasticity: The Brain's Ability to Change

Neuroplasticity is the brain's remarkable ability to adapt and change throughout an individual's life. This ability underpins learning, memory, recovery from brain injury, and the brain's response to new experiences and environments. Unlike the once-prevailing belief that the brain's structure is largely fixed after early development, neuroplasticity shows that the brain is constantly rewiring itself, even into old age.

Structural Neuroplasticity

Structural neuroplasticity refers to the brain's ability to change its physical structure in response to learning and experience. This involves the formation of new connections between neurons (synapses) and the strengthening or weakening of existing ones. When you learn a new skill or form a new memory, the brain undergoes structural changes that make those new behaviors or thoughts easier to repeat in the future.

For example, musicians who practice regularly have been shown to develop larger areas in the brain associated with hand coordination and auditory processing. Similarly, London taxi drivers, who must navigate a complex city, have been found to have an enlarged hippocampus, a region involved in spatial memory and navigation. These examples highlight that the brain physically adapts to the demands placed on it.

Synaptic Plasticity

Synaptic plasticity is the process by which synapses—the connections between neurons—strengthen or weaken over time. This process is crucial for learning and memory. One of the most well-known mechanisms of synaptic plasticity is **long-term potentiation (LTP)**, where repeated stimulation of a synapse increases its strength, making future signals across that synapse more likely to trigger an action potential. LTP is considered a cellular basis for learning and memory.

Conversely, **long-term depression (LTD)** is a process where the strength of synaptic connections decreases, which can be important for forgetting or for refining neural circuits to remove unnecessary connections. Together, LTP and LTD help the brain adapt to new information by modifying the strength of synaptic connections, reinforcing those that are useful and weakening those that are not.

Functional Neuroplasticity

Functional neuroplasticity refers to the brain's ability to move functions from damaged areas to undamaged areas. This is particularly evident after brain injuries, such as strokes, where parts of the brain that are unaffected by the injury can sometimes take over the functions that were lost. For instance, if the left hemisphere of the brain, which typically controls language, is damaged, the right hemisphere may begin to take over some of those language functions.

This type of plasticity is also evident in sensory processing. In individuals who are blind from birth or early in life, the brain's visual cortex, which typically processes visual information, can be repurposed to process non-visual information, such as auditory or tactile stimuli. This repurposing enhances other senses, showing how the brain can adapt to the loss of one sense by strengthening others.

Neuroplasticity in Development and Aging

Neuroplasticity is most pronounced during early development, when the brain is rapidly forming new connections. During this period, the brain is highly responsive to environmental stimuli, allowing children to learn languages, motor skills, and social behaviors at a much faster rate than adults. This critical period of heightened plasticity is crucial for normal development, but it also means that early experiences, whether positive or negative, can have long-lasting effects on brain structure and function.

However, neuroplasticity doesn't stop in childhood. **Throughout adulthood and even into old age, the brain retains the ability to adapt and reorganize**. While plasticity may decline with age, it remains a fundamental feature of the brain. This ongoing plasticity is the basis for lifelong learning and the brain's ability to recover from injuries or adapt to new circumstances, such as learning new technologies or skills later in life.

Neuroplasticity and Mental Health

Neuroplasticity also plays a critical role in mental health. Many psychiatric and neurological disorders, such as depression, anxiety, and post-traumatic stress disorder (PTSD), are associated with maladaptive changes in brain plasticity. For instance, chronic stress can lead to the loss of synapses in the prefrontal cortex, a region involved in decision-making and emotional regulation, while simultaneously increasing connections in the amygdala, which is involved in fear and anxiety.

Understanding neuroplasticity has led to new treatments aimed at promoting positive brain changes. Cognitive-behavioral therapy (CBT), for example, leverages the principles of neuroplasticity to help individuals rewire their thought patterns and behaviors. Similarly, interventions like mindfulness meditation, physical exercise, and certain medications have been shown to encourage beneficial neuroplastic changes, improving mental health and cognitive function.

Neuroplasticity and Rehabilitation

In rehabilitation, particularly after a brain injury or stroke, neuroplasticity is a key concept. Therapeutic interventions aim to harness the brain's plasticity to recover lost functions. For example, **constraint-induced movement therapy (CIMT)** is used to help stroke patients regain movement in a paralyzed limb by encouraging them to use the affected limb instead of compensating with the healthy one. This forced use of the impaired limb promotes plastic changes in the brain, helping to restore motor function.

Overall, neuroplasticity is the brain's dynamic ability to reorganize itself by forming new neural connections. Whether through learning, recovering from injury, or adapting to new environments, neuroplasticity is fundamental to how the brain operates and evolves. Understanding this process provides crucial insights into everything from education and mental health to rehabilitation and aging.

Neurogenesis: Birth of New Neurons

Neurogenesis is the process by which new neurons are formed in the brain. For a long time, the prevailing belief in neuroscience was that the adult brain was incapable of producing new neurons. This view was largely based on the idea that, unlike other cells in the body, neurons in the adult brain were not replaced once lost. However, groundbreaking research in the latter half of the 20th century overturned this notion, revealing that neurogenesis does indeed occur in specific areas of the adult brain.

Where Neurogenesis Occurs

Neurogenesis is most prominently observed in two regions of the brain:

1. **The Hippocampus**: The hippocampus is a critical region for learning, memory, and emotion. Within the hippocampus, neurogenesis takes place in a part called the dentate gyrus. This area is particularly important for the formation of new memories and for the flexibility of thinking, which is essential for adapting to new situations.
2. **The Subventricular Zone (SVZ)**: The SVZ is located along the walls of the lateral ventricles in the brain. In this region, new neurons are generated and then migrate to the olfactory bulb, which is involved in the sense of smell. While the significance of this process in humans is still under investigation, in animals like rodents, it plays a critical role in olfaction.

Stages of Neurogenesis

The process of neurogenesis can be divided into several stages:

1. **Proliferation**: This is the initial stage where neural stem cells, which have the potential to become various types of brain cells, begin to divide. These stem cells are multipotent, meaning they can give rise to neurons as well as glial cells, which support and protect neurons.

2. **Differentiation**: After proliferation, the newly formed cells start to differentiate into neurons or glial cells. If they become neurons, they begin to develop the characteristics of mature neurons, such as axons and dendrites, which are necessary for communication with other neurons.
3. **Migration**: Once differentiated, these young neurons migrate to their destined location within the brain. In the hippocampus, they integrate into existing neural circuits, while in the SVZ, they migrate to the olfactory bulb.
4. **Maturation**: The final stage is maturation, where the new neurons extend their axons and dendrites, form synapses, and become fully functional members of the neural network. This process can take several weeks, and the survival of these new neurons depends on factors like the brain's environment and the individual's activity levels.

Factors Influencing Neurogenesis

Neurogenesis is not a fixed process; it can be influenced by various factors:

- **Exercise**: Physical activity, especially aerobic exercise, has been shown to significantly increase neurogenesis in the hippocampus. Exercise stimulates the production of brain-derived neurotrophic factor (BDNF), a protein that supports the survival and growth of new neurons.
- **Learning and Enrichment**: Engaging in challenging cognitive tasks and living in an enriched environment (with access to social interaction, toys, and novel experiences) also promotes neurogenesis. These activities increase the complexity of the neural environment, encouraging the integration and survival of new neurons.
- **Stress**: Chronic stress, on the other hand, has a negative impact on neurogenesis. High levels of stress hormones, like cortisol, can inhibit the formation of new neurons, particularly in the hippocampus, which may contribute to the development of depression and anxiety disorders.
- **Diet**: Certain dietary factors can influence neurogenesis. For example, diets rich in omega-3 fatty acids, antioxidants, and flavonoids (found in foods like fish, berries, and dark chocolate) are associated with enhanced neurogenesis. Conversely, a high-fat, high-sugar diet can impair neurogenesis.
- **Aging**: Neurogenesis naturally declines with age, which is one reason why cognitive function often diminishes in older adults. However, maintaining an active lifestyle and engaging in mentally stimulating activities can help slow this decline.

Neurogenesis and Mental Health

The discovery of neurogenesis has profound implications for understanding and treating mental health disorders. Reduced neurogenesis in the hippocampus has been linked to depression, anxiety, and cognitive decline. Some antidepressant treatments, such as selective serotonin reuptake inhibitors (SSRIs), have been found to increase neurogenesis, suggesting that part of their therapeutic effect may be due to the promotion of new neuron growth.

Additionally, neurogenesis is thought to play a role in the brain's ability to recover from injury. After a stroke or traumatic brain injury, neurogenesis may help to restore some lost functions by replacing damaged neurons and forming new neural connections.

Controversies and Ongoing Research

While the existence of neurogenesis in the adult brain is widely accepted, there are still debates and ongoing research about its extent and functional significance, especially in humans. Some studies have questioned whether neurogenesis in the adult human hippocampus occurs at levels significant enough to impact cognitive function. This area of research is still evolving, with new discoveries continuing to refine our understanding of neurogenesis.

The Role of Genetics in Neuroscience

Genetics has a critical role in neuroscience, helping us understand how our DNA influences the development, structure, and function of the nervous system. By studying genetics, neuroscientists can unravel the complex relationships between genes, brain function, and behavior. This understanding has profound implications for diagnosing, treating, and potentially preventing neurological and psychiatric disorders.

Genetics and Brain Development

The development of the brain begins early in the embryo, guided by a precise sequence of genetic instructions. These instructions dictate how neural stem cells differentiate into the various types of neurons and glial cells that make up the nervous system. Genes control key processes such as cell proliferation, migration, and differentiation, all of which are essential for building the complex architecture of the brain.

For example, the **HOX genes** play a significant role in the development of the brain and spinal cord by guiding the formation of different body segments during embryogenesis. Mutations in these genes can lead to severe developmental disorders.

Another key player in brain development is the **Notch signaling pathway**, which influences how neural stem cells decide whether to remain stem cells or differentiate into specific types of neurons or glial cells. Disruptions in this pathway can result in abnormal brain development and have been linked to disorders such as autism and schizophrenia.

Genetic Basis of Neurotransmitter Systems

Neurotransmitter systems, which are crucial for communication between neurons, are also under genetic control. Genes encode the enzymes that synthesize neurotransmitters, the receptors that bind them, and the transporters that clear them from the synapse. Variations or mutations in these genes can affect how neurotransmitter systems function, leading to differences in behavior, cognition, and susceptibility to mental health disorders.

For instance, the gene **SLC6A4** encodes the serotonin transporter, which is responsible for reuptaking serotonin from the synapse back into the presynaptic neuron. Variations in this gene have been associated with conditions like depression and anxiety. Individuals with certain variants of the SLC6A4 gene may be more prone to these disorders, particularly under stress.

Genetic Disorders and Neuroscience

Genetics has also been instrumental in identifying the causes of many neurological and psychiatric disorders. **Monogenic disorders**, which are caused by mutations in a single gene, provide clear examples of how genetic alterations can lead to specific brain dysfunctions.

- **Huntington's Disease**: This neurodegenerative disorder is caused by a mutation in the HTT gene, leading to the production of an abnormal version of the huntingtin protein. This abnormal protein gradually damages neurons, particularly in the basal ganglia, leading to symptoms like uncontrolled movements, cognitive decline, and psychiatric problems.
- **Fragile X Syndrome**: This condition is the most common inherited cause of intellectual disability and is linked to a mutation in the FMR1 gene. The mutation prevents the production of a protein that is essential for normal synaptic function, leading to the symptoms of Fragile X Syndrome, which include learning disabilities, social anxiety, and autistic behaviors.
- **Alzheimer's Disease**: While Alzheimer's disease is typically associated with a complex interplay of genetic and environmental factors, mutations in specific genes such as APP, PSEN1, and PSEN2 have been linked to early-onset forms of the disease. These mutations lead to the accumulation of amyloid-beta plaques in the brain, a hallmark of Alzheimer's, which contributes to neurodegeneration and memory loss.

Polygenic Influences and Complex Traits

Most neurological and psychiatric conditions, however, are not caused by a single gene but rather involve the combined effects of many genes—this is known as a **polygenic** influence. Conditions such as schizophrenia, bipolar disorder, and autism spectrum disorders have complex genetic underpinnings where multiple genetic variants contribute to an individual's risk.

For example, **Genome-Wide Association Studies (GWAS)** have identified numerous genetic variants associated with schizophrenia. These studies have found that the risk of schizophrenia is influenced by many small-effect variants across different genes, each contributing a tiny increase in risk. This polygenic nature makes it challenging to predict or diagnose these disorders based solely on genetics, but it also provides a broader understanding of the biological pathways involved.

Epigenetics and Neuroscience

In addition to genetic variation, **epigenetics** is a key concept in neuroscience. Epigenetics refers to changes in gene expression that do not involve alterations to the DNA sequence itself. These changes can be influenced by environmental factors such as stress, diet, and exposure to toxins.

Epigenetic mechanisms include DNA methylation, histone modification, and non-coding RNA regulation, all of which can turn genes on or off without changing the underlying genetic code. For instance, chronic stress can lead to increased DNA methylation of genes involved in the stress response, which may increase vulnerability to depression and anxiety.

Epigenetic changes are particularly important in the brain, where they can influence neural plasticity, learning, and memory. Understanding how these mechanisms work can provide insights into how experiences shape the brain and contribute to the development of mental health disorders.

Personalized Medicine and Genetic Research

As our understanding of the genetic basis of brain function and dysfunction grows, so too does the potential for personalized medicine in neuroscience. By analyzing an individual's genetic makeup, doctors can tailor treatments to their specific needs. For instance, pharmacogenetics—the study of how genes affect a person's response to drugs—can help determine which medications are likely to be most effective or least likely to cause side effects for a particular patient.

In the context of psychiatric disorders, genetic testing might one day help identify individuals who are at higher risk and allow for early interventions that could prevent the onset of illness. Similarly, understanding the genetic basis of neurodegenerative diseases could lead to the development of targeted therapies that slow or halt disease progression.

Challenges and Ethical Considerations

While genetics offers tremendous potential in neuroscience, it also presents significant challenges. The complexity of the brain means that even small genetic changes can have large and sometimes unpredictable effects. Additionally, the ethical implications of genetic research, such as privacy concerns and the potential for genetic discrimination, must be carefully considered.

Understanding the role of genetics in neuroscience is a continually evolving field, providing insights into how our brains develop, function, and sometimes malfunction. This knowledge not only enhances our understanding of the brain but also opens up new possibilities for diagnosing, treating, and preventing a wide range of neurological and psychiatric disorders.

Ethics in Neuroscience

Ethics in neuroscience is a critical area of consideration as advancements in the field raise profound questions about privacy, consent, and the potential misuse of neurotechnologies. One major ethical concern is the privacy of brain data. With the development of neuroimaging and brain-computer interfaces, sensitive information about a person's thoughts, emotions, and intentions could be accessed or even manipulated. Protecting this information from unauthorized use is essential to maintain personal autonomy and dignity.

Informed consent is another key ethical issue. As neuroscience research often involves complex procedures, such as brain scans or invasive techniques, participants must fully understand the potential risks and benefits. Ensuring that participants can make informed decisions about their involvement is crucial, particularly when dealing with vulnerable populations, such as individuals with mental health conditions or cognitive impairments.

The potential for cognitive enhancement through neurotechnology, like deep brain stimulation or smart drugs, also raises ethical concerns. While these technologies could improve the quality of life for individuals with neurological disorders, their use by healthy individuals could lead to issues of fairness and inequality, creating a society where access to cognitive enhancements is determined by wealth or status.

Additionally, the use of neuroscience in legal contexts, such as lie detection or determining criminal responsibility, poses significant ethical challenges. The accuracy and reliability of such techniques are still under scrutiny, and their misuse could result in wrongful convictions or unfair sentencing.

Finally, the ethical implications of animal research in neuroscience must be considered. While animal studies are essential for advancing our understanding of the brain, they must be conducted with the highest standards of care, ensuring that the welfare of animals is respected and that experiments are only conducted when absolutely necessary.

Overall, as neuroscience continues to evolve, it is vital to address these ethical concerns to ensure that the benefits of research and technology are realized without compromising individual rights or societal values.

CHAPTER 2: THE STRUCTURE AND FUNCTION OF NEURONS

Neuronal Anatomy

Neurons are the building blocks of the nervous system, responsible for transmitting information throughout the body. Each neuron is composed of several distinct parts, each with a specific role in its overall function. Understanding the anatomy of a neuron is essential to grasp how these cells communicate and how the brain processes information.

The **cell body**, or soma, is the central part of the neuron. It contains the nucleus, which houses the cell's genetic material. The soma is where most of the neuron's metabolic activities occur, including the production of proteins needed for cell maintenance and the transmission of signals. Surrounding the nucleus are various organelles, like mitochondria, which provide the energy necessary for the neuron's activities.

Branching out from the soma are **dendrites**, which resemble the branches of a tree. Dendrites are responsible for receiving signals from other neurons. These signals, typically in the form of neurotransmitters, bind to receptors on the dendrites' surface, leading to changes in the electrical state of the neuron. The more extensive the dendritic tree, the more information the neuron can receive and process. **Dendritic spines**, small protrusions on the dendrites, further increase the surface area for synaptic connections, making these areas critical for learning and memory.

Extending from the soma in the opposite direction is the **axon**, a long, thin projection that transmits electrical impulses away from the cell body. While dendrites bring information into the neuron, the axon sends it out. The axon can vary in length, from just a fraction of a millimeter to over a meter in some neurons, such as those that run from the spinal cord to the toes. This variation allows neurons to communicate over both short and long distances within the body.

The **axon hillock** is the region where the axon joins the soma. It's a crucial area because it's here that the neuron decides whether or not to send an electrical signal, known as an action potential, down the axon. This decision is based on the summation of all the excitatory and inhibitory signals received by the dendrites. If the combined signals reach a certain threshold, the axon hillock will initiate an action potential.

The axon is often covered with a **myelin sheath**, a fatty layer that insulates the axon and allows electrical impulses to travel more quickly. Myelin is produced by glial cells, specifically oligodendrocytes in the central nervous system and Schwann cells in the peripheral nervous system. The presence of myelin is critical for efficient signal transmission. Without it, the electrical signal would lose strength as it travels down the axon. In myelinated axons, the action potential jumps from one **node of Ranvier** (gaps in the myelin sheath) to the next, a process called saltatory conduction, which speeds up signal transmission.

At the end of the axon are the **axon terminals** or **synaptic boutons**. These structures are responsible for sending the signal to the next neuron or to a target cell, such as a muscle or gland cell. The axon terminals contain vesicles filled with neurotransmitters, the chemical messengers of the nervous system. When an action potential reaches the axon terminals, it triggers the release of these neurotransmitters into the **synapse**, the small gap between the axon terminal of one neuron and the dendrite of the next.

The neurotransmitters then cross the synapse and bind to receptors on the receiving neuron's dendrites, passing the signal along. This process of neurotransmission is how neurons communicate with each other, allowing the brain to perform complex tasks such as thinking, feeling, and controlling movement.

The structure of a neuron is elegantly designed to perform its function—receiving, processing, and transmitting information. The cell body acts as the command center, dendrites as the input receivers, the axon as the transmission line, and the axon terminals as the output deliverers. **Each part of the neuron has a critical role in ensuring that the brain and nervous system can function efficiently and effectively**.

Neurons come in various shapes and sizes, depending on their location and function within the nervous system, but they all share this basic anatomy. This fundamental structure allows them to connect in vast networks, processing and relaying information throughout the body at incredible speeds, making all our thoughts, actions, and sensations possible.

Types of Neurons

Neurons, the fundamental units of the nervous system, come in various types, each specialized for different functions. These types can be broadly categorized based on their structure and function.

Structural Classification
1. **Multipolar Neurons**: These are the most common type of neurons in the central nervous system. They have one axon and multiple dendrites, allowing them to integrate a large amount of information from other neurons. Multipolar neurons are found in the brain and spinal cord and are primarily involved in motor control and higher cognitive functions.

2. **Bipolar Neurons**: These neurons have one axon and one dendrite. They are less common and are primarily found in sensory organs. For instance, bipolar neurons are located in the retina of the eye, where they transmit visual information from photoreceptors to the brain.
   3. **Unipolar Neurons**: Unipolar neurons have a single process extending from the cell body that branches into two directions: one part functions as an axon, and the other part as a dendrite. These neurons are commonly found in the peripheral nervous system and are primarily involved in sensory functions, such as transmitting touch and pain signals from the body to the spinal cord.

Functional Classification
   1. **Sensory Neurons (Afferent Neurons)**: These neurons carry information from sensory receptors (like those in the skin, eyes, and ears) towards the central nervous system. They are vital in perceiving stimuli from the external and internal environment and conveying it to the brain for processing.
   2. **Motor Neurons (Efferent Neurons)**: Motor neurons transmit signals from the central nervous system to muscles and glands, triggering actions such as muscle contraction or glandular secretion. They are essential for voluntary movements as well as involuntary responses like reflexes.
   3. **Interneurons**: Also known as association neurons, interneurons are found exclusively in the central nervous system. They act as connectors, processing information between sensory and motor neurons. Interneurons are involved in complex reflexes and higher cognitive processes like learning and decision-making.

Each type of neuron has a vital role in the functioning of the nervous system, working together to process sensory information, generate responses, and coordinate the body's actions. Their diverse structures and functions are what enable the brain and nervous system to handle the vast array of tasks required for daily life.

Glial Cells

Glial cells, also known as glia, are non-neuronal cells in the nervous system that provide crucial support and protection for neurons. They outnumber neurons by about ten to one and play a variety of essential roles in maintaining brain function.

Types of Glial Cells
   1. **Astrocytes**: Astrocytes are star-shaped cells found in the central nervous system. They perform multiple functions, including maintaining the blood-brain barrier, which protects the brain from harmful substances in the blood. Astrocytes also regulate blood flow to neurons, provide nutrients,

and manage the extracellular environment by removing excess neurotransmitters and ions.
2. **Oligodendrocytes**: Found in the central nervous system, oligodendrocytes are responsible for producing the myelin sheath—a fatty layer that insulates axons and speeds up the transmission of electrical signals. A single oligodendrocyte can extend its processes to multiple axons, myelinating them. In the peripheral nervous system, a similar function is carried out by **Schwann cells**, which myelinate individual axons.
3. **Microglia**: Microglia are the immune cells of the central nervous system. They act as the brain's primary defense mechanism against infections and injury. Microglia continuously monitor the environment for signs of damage or pathogens and respond by engulfing and digesting cellular debris, dead neurons, and pathogens through a process known as phagocytosis.
4. **Ependymal Cells**: These cells line the ventricles of the brain and the central canal of the spinal cord. Ependymal cells are involved in the production and circulation of cerebrospinal fluid (CSF), which cushions the brain and spinal cord, removing waste and transporting nutrients.

Functions of Glial Cells

Glial cells are not just passive support cells; they actively contribute to brain function. They regulate the extracellular environment, maintain homeostasis, and provide structural support to neurons. Glial cells also play a role in synaptic plasticity, the process by which synapses strengthen or weaken, which is essential for learning and memory. Additionally, they are involved in the repair and scarring process following brain injury.

In sum, glial cells are indispensable to the health and function of the nervous system. They support neurons in numerous ways, ensuring that the brain operates smoothly and efficiently. Without glial cells, neurons would not be able to perform their functions effectively, highlighting the critical role these cells play in brain health and disease.

Neuronal Communication

Neuronal communication is the process by which neurons transmit information throughout the nervous system, enabling everything from basic reflexes to complex thoughts. This communication occurs both electrically and chemically, ensuring rapid and precise signaling.

Electrical Communication

The process begins with an **action potential**, an electrical impulse generated when a neuron is sufficiently stimulated. This occurs at the axon hillock, where the neuron decides to fire based on incoming signals from other neurons. If the combined input reaches a threshold, voltage-gated ion channels open, allowing positively charged sodium ions to rush into the neuron. This influx of ions causes a rapid change in the electrical charge of the neuron, creating the action potential.

The action potential travels down the axon to the axon terminals. In myelinated neurons, this transmission is sped up by the myelin sheath, which allows the action potential to jump from one node of Ranvier to the next, a process known as saltatory conduction.

Chemical Communication

When the action potential reaches the axon terminals, it triggers the release of **neurotransmitters**—chemical messengers stored in vesicles. These neurotransmitters are released into the **synapse**, the small gap between the sending neuron's axon terminal and the receiving neuron's dendrite.

The neurotransmitters cross the synapse and bind to specific receptors on the receiving neuron's surface. This binding changes the electrical state of the receiving neuron, either exciting it (making it more likely to fire its own action potential) or inhibiting it (making it less likely to fire). The type of neurotransmitter and receptor involved determines the nature of the response.

After the neurotransmitter has done its job, it must be cleared from the synapse to prevent continuous stimulation. This is achieved through reuptake (where the neurotransmitter is taken back into the presynaptic neuron), enzymatic breakdown, or diffusion away from the synapse.

Integration of Signals

Neurons receive inputs from many other neurons, and these signals can be excitatory or inhibitory. The neuron integrates these signals at the axon hillock. If the excitatory signals outweigh the inhibitory ones and the threshold is reached, the neuron fires an action potential, continuing the cycle of communication.

This precise and coordinated communication between neurons is what underlies all brain functions, from simple reflexes to complex cognitive processes, making it the foundation of the nervous system's operation.

Synaptic Structure and Function

The synapse is a crucial structure in the nervous system, where communication between neurons occurs. It is the point of contact where the axon terminal of one neuron meets the dendrite or cell body of another, allowing the transfer of information. The synapse is essential for transmitting signals that underlie all brain functions.

Structure of the Synapse

A synapse consists of three main components:

1. **Presynaptic Terminal**: This is the end of the axon from the sending neuron. It contains synaptic vesicles filled with neurotransmitters, the chemical messengers used in neuronal communication.
2. **Synaptic Cleft**: This is the small gap between the presynaptic terminal and the receiving neuron, typically around 20-40 nanometers wide. Despite its small size, it's crucial for ensuring that the neurotransmitter must travel across to reach the postsynaptic neuron, allowing for precise control of signal transmission.
3. **Postsynaptic Membrane**: Located on the dendrite or cell body of the receiving neuron, this membrane contains specific receptors that bind to neurotransmitters. These receptors are proteins that detect and respond to the chemical signals released by the presynaptic neuron.

Function of the Synapse

When an action potential reaches the presynaptic terminal, it triggers the opening of voltage-gated calcium channels. The influx of calcium ions causes synaptic vesicles to fuse with the presynaptic membrane, releasing neurotransmitters into the synaptic cleft.

The neurotransmitters then diffuse across the cleft and bind to receptors on the postsynaptic membrane. This binding opens ion channels in the postsynaptic neuron, leading to changes in its electrical state. Depending on the type of neurotransmitter and receptor, this can result in an excitatory or inhibitory response.

If the signal is excitatory, the postsynaptic neuron becomes more likely to fire its own action potential. If inhibitory, the neuron becomes less likely to fire. This process is how neurons communicate, allowing for complex processing of information.

After the neurotransmitter has acted, it is either reabsorbed by the presynaptic neuron (reuptake), broken down by enzymes, or diffused away from the synapse. This ensures that the synapse is ready for the next signal.

Synaptic structure and function are central to the brain's ability to process and transmit information, making them fundamental to all aspects of nervous system activity.

Neuronal Networks and Circuits

Neuronal networks and circuits are the intricate systems that enable the brain to process information, perform complex tasks, and regulate bodily functions. These networks consist of interconnected neurons that communicate with one another through synapses, forming the basis of all neural activity.

Neuronal Networks

A neuronal network is a group of neurons that are interconnected and work together to perform specific functions. These networks can range from simple reflex arcs to complex circuits involved in higher-order cognitive processes. In a basic sense, networks consist of **sensory neurons** that receive input from the environment, **interneurons** that process this information, and **motor neurons** that execute a response.

In more complex networks, such as those involved in cognition or emotion, thousands or even millions of neurons can be interconnected. These networks are highly dynamic, meaning they can change and adapt in response to new information, learning, and experience. **Synaptic plasticity**—the ability of synapses to strengthen or weaken over time—is a key mechanism that allows neuronal networks to adapt and form the basis of memory and learning.

Neuronal Circuits

Neuronal circuits are specific pathways within neuronal networks that are responsible for particular functions. A circuit typically involves a sequence of neurons that are connected in a precise manner to carry out a specific task. For example, in a simple reflex circuit, sensory neurons detect a stimulus, such as a tap on the knee, and directly connect to motor neurons in the spinal cord, which then cause the leg to kick. This is a straightforward, linear circuit.

More complex circuits, like those in the cerebral cortex, involve multiple layers of neurons and feedback loops. **Recurrent circuits**, where neurons loop back to earlier neurons in the pathway, are common in processes such as memory retrieval and decision-making. **Parallel processing circuits** allow the brain to process different aspects of information simultaneously, such as recognizing a face while also processing the emotions it conveys.

Integration and Coordination

Neuronal networks and circuits do not operate in isolation. They are integrated and coordinated across different regions of the brain and body. For example, the visual system involves multiple circuits that process color, motion, and depth, all of which are integrated to produce a cohesive visual experience. Similarly, circuits in the limbic system integrate sensory information with emotional responses, guiding behavior in a contextually appropriate manner.

These networks and circuits are highly specialized and organized, allowing for the precise control and regulation of various physiological and cognitive processes. For instance, the brain's ability to perform tasks such as speech, movement, and decision-making relies on the seamless integration of multiple neuronal circuits that span different brain regions. The **prefrontal cortex**, **basal ganglia**, and **cerebellum** work in concert to plan, initiate, and fine-tune motor actions, ensuring smooth and coordinated movement.

Additionally, **feedforward** and **feedback mechanisms** are essential in refining neural responses. Feedforward pathways allow information to flow from one neuron or network to the next, driving immediate responses. Feedback pathways, on the other hand, allow later stages of processing to influence earlier ones, providing a system of checks and balances that ensures accuracy and adaptability in response to new stimuli or changing conditions.

Neuronal oscillations, or rhythmic patterns of neural activity, are another critical feature of these networks, facilitating synchronization across different parts of the brain. These oscillations allow for coordinated activity in distant neuronal circuits, essential for functions like attention, perception, and consciousness. Disruptions in these oscillatory patterns are associated with various neurological and psychiatric disorders, underscoring the importance of the balance and timing within neuronal networks and circuits.

CHAPTER 3: NEUROTRANSMITTERS AND THEIR RECEPTORS

Major Neurotransmitters

Neurotransmitters are the chemical messengers that neurons use to communicate with each other and with other cells in the body. These molecules play critical roles in regulating a wide range of functions, from mood and cognition to muscle contraction. Here are some of the major neurotransmitters and their functions:

Glutamate

Glutamate is the most common excitatory neurotransmitter in the brain. It's crucial for synaptic plasticity, which underlies learning and memory. When glutamate binds to its receptors, such as NMDA and AMPA receptors, it increases the likelihood that the receiving neuron will fire an action potential. However, too much glutamate can be harmful, leading to excitotoxicity, which is associated with conditions like stroke and neurodegenerative diseases.

GABA (Gamma-Aminobutyric Acid)

GABA is the primary inhibitory neurotransmitter in the brain. It works by binding to GABA receptors, which open channels that allow negatively charged ions to enter the neuron, making it less likely to fire an action potential. GABA is essential for maintaining the brain's balance between excitation and inhibition, preventing excessive neuronal activity that could lead to seizures. Many anti-anxiety medications and sedatives work by enhancing GABA's effects.

Dopamine

Dopamine is involved in several important brain functions, including motivation, reward, and motor control. It's released in areas of the brain like the basal ganglia and the prefrontal cortex. Dopamine is crucial for reinforcing behaviors that are pleasurable or rewarding. Dysfunction in dopamine signaling is linked to disorders such as Parkinson's disease, where dopamine-producing neurons degenerate, and schizophrenia, which is associated with altered dopamine pathways.

Serotonin

Serotonin is a neurotransmitter that affects mood, appetite, sleep, and cognition. It's produced in the brainstem and is widely distributed throughout the brain. Serotonin binds to a variety of receptors, each influencing different physiological processes. For example, the serotonin 5-HT1A receptor is involved in mood regulation, and imbalances in serotonin levels are linked to depression and anxiety. Selective serotonin reuptake inhibitors (SSRIs), a common class of antidepressants, work by increasing the availability of serotonin in the synapse.

Acetylcholine

Acetylcholine plays a key role in both the central and peripheral nervous systems. In the brain, it's involved in attention, learning, and memory. In the peripheral nervous system, it's the neurotransmitter that triggers muscle contraction. Acetylcholine binds to two main types of receptors: nicotinic and muscarinic. The loss of acetylcholine-producing neurons is a hallmark of Alzheimer's disease, leading to cognitive decline.

Norepinephrine

Norepinephrine functions as both a neurotransmitter and a hormone. It's involved in the body's "fight or flight" response, increasing heart rate, blood pressure, and blood flow to muscles. In the brain, norepinephrine is important for attention, arousal, and alertness. It's produced in the locus coeruleus and released throughout the brain, influencing various cognitive and physiological processes.

These major neurotransmitters are essential for the brain's ability to process information and control the body. Each one plays a unique role, contributing to the complex web of interactions that underlie all aspects of behavior and cognition. Understanding how they work provides insight into both normal brain function and the basis of various neurological and psychiatric disorders.

Receptor Types

Neurotransmitter receptors are proteins located on the surface of neurons and other cells, where they are important in the communication process by binding to specific neurotransmitters. These receptors are generally categorized into two main types: ionotropic and metabotropic.

Ionotropic Receptors

Ionotropic receptors are also known as ligand-gated ion channels. When a neurotransmitter binds to an ionotropic receptor, the receptor directly opens an ion channel within the cell membrane. This allows specific ions, such as sodium (Na+), potassium (K+), or chloride (Cl-), to flow in or out of the cell. The flow of these ions changes the cell's electrical state, which can either excite or inhibit the neuron.

- **Excitatory Ionotropic Receptors**: These include receptors like the **AMPA** and **NMDA receptors** for glutamate. When activated, they allow positively charged ions to enter the neuron, making it more likely to fire an action potential. This excitatory process is essential for fast synaptic transmission and plays a key role in learning and memory.

- **Inhibitory Ionotropic Receptors**: The **GABA-A receptor** is an example of an inhibitory ionotropic receptor. When GABA binds to this receptor, it opens a channel that allows negatively charged chloride ions to enter the neuron. This influx makes the neuron less likely to fire, helping to regulate and calm neuronal activity.

Metabotropic Receptors

Metabotropic receptors work differently. They do not contain an ion channel but instead are coupled to G-proteins, which initiate a cascade of intracellular events when a neurotransmitter binds to the receptor. These events can lead to changes in ion channels elsewhere on the cell membrane, alter gene expression, or modify other cellular processes.

- **G-Protein-Coupled Receptors (GPCRs)**: These are the most common type of metabotropic receptors. For instance, the **serotonin 5-HT1A receptor** is a GPCR that, when activated, can influence mood by modulating neurotransmitter release and intracellular signaling pathways. The effects of metabotropic receptors are typically slower and longer-lasting compared to ionotropic receptors, making them crucial for regulating complex behaviors like mood, appetite, and cognition.
- **Muscarinic Acetylcholine Receptors**: Another example of metabotropic receptors is the muscarinic receptors for acetylcholine, which play significant roles in the central nervous system (like in memory and learning) and the peripheral nervous system (like in controlling heart rate and smooth muscle contraction).

Ionotropic receptors are fast-acting, providing immediate responses, while metabotropic receptors are slower but more versatile, influencing a wide range of cellular functions. Both types of receptors are essential for the proper functioning of the nervous system, enabling precise control over neural communication and responses.

Neurotransmitter Systems and Behavior

Neurotransmitter systems play a pivotal role in shaping behavior by influencing how neurons communicate across different regions of the brain. Each neurotransmitter system is linked to specific behaviors and mental processes, and imbalances or disruptions in these systems can lead to various psychological and neurological conditions.

Dopamine System

The dopamine system is integral to reward, motivation, and pleasure. Dopamine neurons are concentrated in areas like the substantia nigra and the ventral tegmental area (VTA). From there, dopamine pathways project to various parts of the brain, including the prefrontal cortex and the nucleus accumbens. This system underlies behaviors related to reward-seeking and reinforcement learning. Dysfunction in the dopamine system is linked to disorders such as Parkinson's disease, where dopamine-producing neurons degenerate, and schizophrenia, where dopamine dysregulation contributes to symptoms like hallucinations and delusions.

Serotonin System

The serotonin system influences mood, appetite, sleep, and emotional regulation. Serotonin is primarily produced in the raphe nuclei of the brainstem and is widely distributed throughout the brain. It modulates mood and anxiety levels, which is why selective serotonin reuptake inhibitors (SSRIs) are commonly used to treat depression and anxiety disorders. Low levels of serotonin are associated with mood disorders, while excess levels can contribute to conditions like serotonin syndrome.

GABA and Glutamate Systems

GABA (gamma-aminobutyric acid) and glutamate are the primary inhibitory and excitatory neurotransmitters, respectively. **The GABA system** helps regulate anxiety and stress by calming neuronal activity. Drugs that enhance GABA activity, like benzodiazepines, are used to treat anxiety and epilepsy. **The glutamate system** is essential for learning and memory, as it promotes synaptic plasticity. However, excessive glutamate activity can lead to excitotoxicity, contributing to neurodegenerative diseases like Alzheimer's and Huntington's.

Norepinephrine System

Norepinephrine, produced in the locus coeruleus, is important in the "fight or flight" response, attention, and arousal. It prepares the body for action in stressful situations by increasing heart rate and blood flow to muscles. Dysregulation of the norepinephrine system is linked to conditions like anxiety disorders and PTSD, where the system's heightened activity can lead to hypervigilance and an exaggerated stress response.

These neurotransmitter systems are intricately linked to behavior, influencing everything from basic survival instincts to complex emotional states. Understanding these systems provides insight into how the brain controls behavior and how imbalances can lead to mental health disorders.

Pharmacology of Neurotransmission

Pharmacology of neurotransmission involves the study of how drugs affect the communication between neurons by interacting with neurotransmitter systems. This interaction can either enhance or inhibit the normal processes of neurotransmission, leading to therapeutic effects or side effects, depending on the drug and its target.

Agonists and Antagonists

Drugs can act as **agonists** or **antagonists** at neurotransmitter receptors. **Agonists** bind to receptors and mimic the action of a neurotransmitter, enhancing its effects. For example, opioid drugs like morphine act as agonists at opioid receptors, mimicking endorphins to relieve pain. **Antagonists**, on the other hand, block receptors and prevent neurotransmitters from exerting their effects. An example is naloxone, which acts as an antagonist at opioid receptors, used to reverse opioid overdoses by blocking the effects of the opioid agonists.

Reuptake Inhibitors

Some drugs affect neurotransmitter levels by inhibiting their reuptake into the presynaptic neuron. **Selective serotonin reuptake inhibitors (SSRIs)**, commonly used to treat depression, work by blocking the reuptake of serotonin, increasing its availability in the synaptic cleft and prolonging its action on the postsynaptic neuron. This mechanism helps alleviate symptoms of depression by enhancing serotonin signaling.

Enzyme Inhibitors

Enzyme inhibitors work by preventing the breakdown of neurotransmitters. **Monoamine oxidase inhibitors (MAOIs)**, used to treat depression, block the enzyme monoamine oxidase, which normally breaks down neurotransmitters like serotonin, dopamine, and norepinephrine. By inhibiting this enzyme, MAOIs increase the levels of these neurotransmitters, thereby boosting mood and alleviating depressive symptoms.

Modulation of Ion Channels

Certain drugs modulate ion channels directly, affecting neuronal excitability. **Benzodiazepines**, for instance, enhance the effect of the inhibitory neurotransmitter GABA by increasing the opening of GABA-A receptor channels, allowing more chloride ions to enter the neuron. This hyperpolarizes the neuron, making it less likely to fire and producing a calming effect, which is useful in treating anxiety and epilepsy.

Understanding the pharmacology of neurotransmission is crucial for developing medications that can precisely target specific neurotransmitter systems, offering therapeutic benefits while minimizing side effects.

Neuromodulators and Their Effects

Neuromodulators are a class of neurotransmitters that regulate the activity of neural circuits by altering the strength or efficacy of synaptic transmission. Unlike classical neurotransmitters, which typically have immediate and localized effects, neuromodulators influence a broader range of neurons over longer periods, shaping overall brain function and behavior.

Dopamine

Dopamine acts as both a neurotransmitter and a neuromodulator. As a neuromodulator, it plays a critical role in reward processing, motivation, and motor control. Dopamine's modulatory effects are crucial for reinforcing behaviors associated with pleasure and reward. Dysregulation of dopamine can lead to disorders like Parkinson's disease, characterized by motor deficits due to the loss of dopamine-producing neurons, and addiction, where dopamine pathways are hijacked by substances or behaviors that cause excessive dopamine release.

Serotonin

Serotonin is another key neuromodulator, widely involved in mood regulation, anxiety, and appetite. It modulates various neural circuits by influencing the activity of multiple neurotransmitter systems. For example, serotonin can enhance or inhibit the release of other neurotransmitters, depending on the receptors it binds to. This broad influence helps maintain emotional stability and plays a role in conditions like depression and anxiety, where serotonin levels or receptor function are often disrupted.

Norepinephrine

Norepinephrine, produced primarily in the locus coeruleus, acts as a neuromodulator by affecting arousal, attention, and the body's stress response. It enhances the responsiveness of neural circuits involved in alertness and vigilance, preparing the brain to respond to challenging situations. Elevated norepinephrine levels are associated with heightened alertness and readiness, while dysregulation can contribute to anxiety disorders and hypervigilance, as seen in PTSD.

Acetylcholine

Acetylcholine serves as a neuromodulator in both the central and peripheral nervous systems. In the brain, it is involved in learning, memory, and attention. Acetylcholine modulates neural circuits by altering the activity of other neurotransmitters and is crucial for cognitive functions. Dysfunction in cholinergic systems is linked to memory impairments in diseases like Alzheimer's.

Neuromodulators have a vital role in fine-tuning the brain's activity, affecting large networks of neurons and influencing complex behaviors and physiological states. Their broad and prolonged effects make them essential for maintaining the brain's overall balance and function.

Neurotransmitter Synthesis, Degradation, Transport, and Reuptake

Neurotransmitters are the chemical messengers of the nervous system, and their synthesis, degradation, transport, and reuptake are crucial for proper neural communication. Each of these processes is tightly regulated to ensure that neurotransmission is efficient, precise, and capable of adapting to the brain's needs.

Synthesis

Neurotransmitters are synthesized from precursor molecules, typically amino acids or other small organic compounds, within the neuron. The synthesis process is catalyzed by specific enzymes. For example, **dopamine** is synthesized from the amino acid tyrosine, which is converted into L-DOPA by the enzyme tyrosine hydroxylase. L-DOPA is then converted into dopamine by the enzyme DOPA decarboxylase. Similarly, **serotonin** is synthesized from the amino acid tryptophan, which is converted to 5-hydroxytryptophan (5-HTP) and then to serotonin by tryptophan hydroxylase and aromatic L-amino acid decarboxylase, respectively. Once synthesized, neurotransmitters are stored in synaptic vesicles within the presynaptic terminal, ready for release.

Degradation

Neurotransmitter degradation occurs primarily after the neurotransmitter has been released into the synapse and has bound to its receptors on the postsynaptic neuron. Enzymes specific to each neurotransmitter break them down to terminate their action and prevent continuous stimulation. For example, **acetylcholine** is broken down by the enzyme acetylcholinesterase into acetate and choline, effectively terminating the signal. Similarly, **monoamine oxidase (MAO)** and **catechol-O-methyltransferase (COMT)** are enzymes involved in the breakdown of catecholamines like dopamine, norepinephrine, and serotonin.

Transport

Transport refers to the movement of neurotransmitters to and from various locations within the neuron. After synthesis in the neuron's cell body, neurotransmitters are transported down the axon to the presynaptic terminal in vesicles via axonal transport. Vesicles containing neurotransmitters are then transported to the synaptic cleft during neurotransmission. In addition, some neurotransmitters are synthesized directly in the presynaptic terminal.

Reuptake

Reuptake is the process by which neurotransmitters are reabsorbed into the presynaptic neuron after they have performed their function. Specialized transporter proteins in the presynaptic membrane carry out this process. For example, **serotonin** is taken back up by the serotonin transporter (SERT), and

dopamine by the dopamine transporter (DAT). Reuptake is crucial for recycling neurotransmitters for future use and for clearing the synaptic cleft, thus stopping the neurotransmitter's action and preventing overstimulation of the postsynaptic neuron.

Reuptake inhibitors, such as SSRIs (Selective Serotonin Reuptake Inhibitors) for serotonin, are commonly used in treating depression by blocking the reuptake process, thereby increasing the availability of the neurotransmitter in the synapse and prolonging its action.

These processes—synthesis, degradation, transport, and reuptake—work in concert to regulate neurotransmitter levels and ensure precise control over neural communication. Any disruption in these processes can lead to neurological and psychiatric disorders, highlighting their importance in maintaining brain health.

CHAPTER 4: THE DEVELOPMENT OF THE NERVOUS SYSTEM

Embryonic Development

Embryonic development of the nervous system begins early in pregnancy, laying the foundation for the brain, spinal cord, and peripheral nerves. This process, called **neurulation**, starts just a few weeks after fertilization and is highly orchestrated, with each step building on the previous one.

Around the third week of gestation, a specialized layer of cells in the embryo known as the **neural plate** forms. This plate is the origin of the entire nervous system. The cells in the neural plate undergo a series of changes, thickening and folding inwards to form the **neural groove**. As the groove deepens, the edges, or neural folds, move towards each other and eventually fuse, creating the **neural tube**.

The neural tube is a crucial structure, as it will develop into the central nervous system. The anterior (head) end of the neural tube enlarges and differentiates into three primary brain vesicles: the **forebrain (prosencephalon)**, **midbrain (mesencephalon)**, and **hindbrain (rhombencephalon)**. These vesicles will give rise to major brain regions, including the cerebral cortex, thalamus, brainstem, and cerebellum. The posterior (tail) end of the neural tube extends to form the spinal cord.

Neural crest cells, a group of cells that emerge from the edges of the neural tube, migrate throughout the embryo to form various structures, including the peripheral nervous system, which connects the central nervous system to the rest of the body. These cells differentiate into sensory neurons, autonomic neurons, and glial cells, as well as non-neural structures like facial cartilage and the adrenal medulla.

As the neural tube closes, the cells within it begin to proliferate rapidly, producing the precursors to neurons and glial cells. These precursor cells, known as **neuroblasts** and **glioblasts**, migrate to their appropriate locations, guided by chemical signals and physical cues within the developing embryo. Once they reach their destination, they differentiate into specific types of neurons or glia, forming the intricate networks that will later support brain function.

During this period, the brain undergoes extensive growth and folding, particularly in the cerebral cortex, which is responsible for higher-order functions like thought, memory, and emotion. **Synaptogenesis**, the formation of synapses between neurons, begins in the embryonic stage and continues after birth, allowing for the complex communication necessary for brain activity.

The precise timing and regulation of these processes are critical. Disruptions can lead to congenital conditions such as spina bifida, where the neural tube fails to close completely, or microcephaly, where brain growth is significantly impaired.

Embryonic development of the nervous system is a highly dynamic and intricate process that sets the stage for all subsequent neural development, ultimately determining the functional architecture of the brain and nervous system.

Cell Migration and Axon Guidance

During the development of the nervous system, **cell migration** and **axon guidance** are crucial processes that shape the brain and spinal cord, ensuring that neurons reach their correct positions and form the appropriate connections.

Cell Migration

Cell migration begins after neural progenitor cells, known as neuroblasts, are generated in specific regions of the developing brain, such as the ventricular zone. These neuroblasts need to move to their final destinations, where they will differentiate into various types of neurons. This migration is guided by a combination of signals, both chemical and physical, that direct the cells along specific pathways.

There are several modes of cell migration, but one of the most common in the developing brain is **radial migration**. In radial migration, neuroblasts move along radial glial fibers, which extend from the ventricular zone to the outer surface of the brain. These fibers act as scaffolds, guiding the neuroblasts to their target locations in the cortical layers. Proper migration is essential for the formation of the brain's layered structure, particularly in the cerebral cortex, where neurons must be precisely positioned to form functional circuits.

Disruptions in cell migration can lead to developmental disorders such as lissencephaly, where the brain's surface appears smooth due to improper layering of neurons.

Axon Guidance

Once neurons have reached their final locations, they begin to extend axons, the long projections that will connect them to other neurons, muscles, or glands. **Axon guidance** is the process by which these axons find their way to their appropriate targets, forming the intricate wiring of the nervous system.

Axon guidance is directed by a complex set of cues in the embryonic environment. These cues can be **attractive or repulsive**, steering the growing axon toward or away from certain areas. For example, **netrins** are a family of proteins that can attract axons, guiding them along specific pathways, while **semaphorins** serve as repulsive signals, preventing axons from entering certain regions.

The growth cone, a dynamic structure at the tip of the extending axon, is used in sensing these guidance cues. The growth cone responds to the molecular environment, making decisions about which direction to grow. It extends filopodia, thin protrusions that explore the surrounding area, allowing the axon to navigate through complex environments to reach its target.

Once the axon reaches its target, it forms synapses, establishing the connections that will enable communication within the nervous system. The precision of axon guidance is vital for proper neural circuit formation; errors in this process can lead to neurological disorders, such as congenital mirror movement disorder, where incorrect wiring leads to involuntary mirror movements on both sides of the body.

Cell migration and axon guidance work in tandem to create the intricate architecture of the nervous system, ensuring that neurons are properly placed and connected to form functional neural circuits. These processes are essential for the brain's ability to process information and control behavior.

Synaptogenesis

Synaptogenesis is the process by which synapses, the connections between neurons, are formed. This process is crucial for the development of the nervous system, as it establishes the networks that allow neurons to communicate and process information.

Synaptogenesis begins during early brain development and continues into adulthood, though the most intense period occurs in the first few years of life. During this time, neurons extend their axons and dendrites, seeking out other neurons to connect with. When an axon reaches its target, it forms a synapse with the dendrite or cell body of the receiving neuron.

The formation of synapses is highly regulated by both genetic and environmental factors. Specific molecular signals guide the growth cone of the axon to its target, where it releases neurotransmitters that help establish the connection. These signals include cell adhesion molecules, which physically anchor the axon to the target, and trophic factors, which support the growth and survival of the synapse.

Once a synapse is formed, it must be stabilized and refined. **Activity-dependent processes** play a significant role in this stage. Synapses that are frequently used become stronger and more stable, while those that are less active may weaken and eventually be eliminated, a process known as synaptic pruning. This refinement ensures that the neural circuits that remain are the most efficient and effective for processing information.

Synaptogenesis is critical for learning and memory. The brain's ability to adapt and reorganize in response to experience, known as neuroplasticity, relies on the formation of new synapses. This plasticity is particularly evident in early childhood, a period marked by rapid synapse formation, but it continues throughout life, enabling the brain to learn new skills and recover from injuries.

Disruptions in synaptogenesis are associated with various neurological and psychiatric disorders. For example, abnormal synapse formation and pruning have been linked to autism spectrum disorders and schizophrenia, highlighting the importance of precise synaptic development for normal brain function.

Overall, synaptogenesis is a fundamental process that underlies the brain's ability to form functional networks, essential for all aspects of cognition and behavior.

Critical Periods in Development

Critical periods are specific windows during development when the brain is particularly receptive to certain environmental stimuli. During these periods, the nervous system undergoes rapid changes that are essential for the proper formation of neural circuits and the acquisition of specific skills and behaviors. Once a critical period ends, the brain's plasticity in that particular domain diminishes, making it more difficult to acquire certain abilities or make up for missed experiences.

Visual Development

One of the most well-studied examples of a critical period is in visual development. During early infancy, the brain is highly sensitive to visual input. If a child's visual experience is disrupted during this time—such as by cataracts or strabismus (misaligned eyes)—the visual cortex may not develop properly. This can lead to permanent deficits in vision, known as amblyopia, even if the physical issue is later corrected. The critical period for visual development typically occurs within the first few years of life.

Language Acquisition

Language development is another area where critical periods are evident. Children are most adept at learning languages during the early years, roughly from birth to around age seven. During this time, they can easily pick up multiple languages, achieving native-like fluency. As they grow older, the brain becomes less plastic in terms of language learning, and acquiring a new language becomes more challenging and often less successful in terms of accent and grammatical fluency.

Social and Emotional Development

Social and emotional development also has critical periods, particularly in early childhood. Attachment between a caregiver and a child, formed during the first few years, is crucial for healthy emotional development. Disruptions in this attachment process, such as in cases of severe neglect or abuse, can lead to long-term emotional and social difficulties, including attachment disorders and impaired social functioning.

Cognitive Skills

Cognitive skills, such as problem-solving and reasoning, also develop during critical periods. The early years are crucial for the development of executive functions, which include working memory, cognitive flexibility, and inhibitory control. Environments rich in stimuli and learning opportunities during these periods can significantly enhance cognitive abilities.

Critical periods are essential for the brain's development, ensuring that neural circuits are shaped by appropriate experiences. Missing these windows can lead to lasting deficits, highlighting the importance of early intervention and enriched environments during these key stages of development.

Developmental Disorders of the Nervous System

Developmental disorders of the nervous system are conditions that arise during the formation and maturation of the brain and spinal cord, leading to long-term impacts on cognitive, motor, and emotional functions. These disorders can result from genetic mutations, environmental factors, or a combination of both, affecting the nervous system at various stages of development.

Autism Spectrum Disorder (ASD)

Autism Spectrum Disorder is a neurodevelopmental condition characterized by challenges in social interaction, communication, and repetitive behaviors. The exact causes of ASD are complex and involve both genetic and environmental factors. During early brain development, abnormalities in synaptogenesis, neural connectivity, and brain growth are thought to contribute to the symptoms of autism. These disruptions can affect how neurons communicate, leading to the characteristic behaviors and difficulties associated with the disorder.

Attention-Deficit/Hyperactivity Disorder (ADHD)

ADHD is a developmental disorder marked by inattention, hyperactivity, and impulsivity. It is believed to involve delayed maturation of the prefrontal cortex, the brain region responsible for executive functions like attention, planning, and impulse control. Genetic factors play a significant role in ADHD, and environmental influences, such as prenatal exposure to toxins, may also contribute. Disruptions in the dopaminergic and noradrenergic systems, which regulate attention and arousal, are also implicated in ADHD.

Cerebral Palsy

Cerebral palsy is a group of disorders that affect movement, muscle tone, and posture. It results from damage to the developing brain, often before birth, but sometimes during delivery or early infancy. Causes include oxygen deprivation (hypoxia), infections, or traumatic injury. The damage primarily affects the motor cortex, basal ganglia, and cerebellum, leading to difficulties in controlling movement and coordination. The severity of symptoms can vary widely, from mild motor impairments to significant physical disabilities.

Intellectual Disabilities

Intellectual disabilities involve limitations in intellectual functioning and adaptive behavior. These disabilities can result from various factors, including genetic conditions like Down syndrome, where an extra chromosome 21 causes developmental delays and intellectual impairment. Environmental factors such as malnutrition, exposure to toxins, or infections during pregnancy can also contribute. These conditions impact the development of neural circuits involved in learning, memory, and problem-solving.

Fragile X Syndrome

Fragile X Syndrome is a genetic disorder caused by a mutation in the FMR1 gene, leading to intellectual disability, behavioral challenges, and often, features of autism. The mutation disrupts the production of a protein essential for synaptic development and plasticity. This affects how neurons form connections, leading to the cognitive and behavioral symptoms observed in the disorder.

Spina Bifida

Spina bifida is a neural tube defect that occurs when the neural tube fails to close properly during early embryonic development. This results in the incomplete formation of the spinal cord and vertebrae, leading to physical disabilities that range from mild to severe. Depending on the severity, individuals with spina bifida may experience mobility issues, bladder and bowel dysfunction, and, in some cases, intellectual challenges.

Developmental Coordination Disorder (DCD)

DCD, also known as dyspraxia, affects motor coordination and planning. Children with DCD struggle with tasks that require fine and gross motor skills, such as writing, tying shoelaces, or participating in sports. The exact causes are not fully understood, but it is believed to involve disruptions in the brain areas responsible for motor control, such as the cerebellum and parietal lobe. DCD often co-occurs with other developmental disorders, such as ADHD or learning disabilities.

Developmental disorders of the nervous system vary widely in their presentation and severity, but all share a common feature: they result from disruptions to the normal development and functioning of the brain and spinal cord. Early diagnosis and intervention can significantly improve outcomes, helping individuals achieve their full potential despite these challenges.

CHAPTER 5: SENSORY SYSTEMS: HOW WE PERCEIVE THE WORLD

The Visual System

The visual system is our primary way of perceiving the world, converting light into electrical signals that the brain interprets as images. This process begins when light enters the eye through the **cornea**, the transparent outer layer that helps focus light. The light then passes through the **pupil**, an adjustable opening controlled by the **iris**, the colored part of the eye. The iris regulates the amount of light entering the eye by expanding or contracting the pupil.

Once through the pupil, light is focused by the **lens** onto the **retina**, a thin layer of tissue at the back of the eye. The lens adjusts its shape to focus light properly, a process called accommodation. The retina contains millions of photoreceptor cells, specifically **rods** and **cones**, which are responsible for detecting light.

Rods are more numerous and are highly sensitive to low light levels, making them essential for night vision. They don't detect color but are excellent at perceiving shapes and motion in dim conditions. **Cones**, on the other hand, are responsible for color vision and are concentrated in the **fovea**, the central part of the retina. Cones require brighter light to function and allow us to see fine details and colors.

When light hits the photoreceptors, it triggers a chemical reaction that changes the shape of a molecule called **rhodopsin** in rods or **photopsin** in cones. This change starts a cascade of events that generate an electrical signal. These signals are processed by other cells in the retina, such as bipolar and ganglion cells, before being transmitted to the brain.

The axons of the **ganglion cells** converge to form the **optic nerve**, which carries visual information from the retina to the brain. The optic nerve from each eye meets at the **optic chiasm**, where fibers from the inner halves of the retina cross over to the opposite side of the brain. This crossing ensures that visual information from the right visual field is processed in the left hemisphere of the brain and vice versa.

After the optic chiasm, the signals travel through the **optic tracts** to the **lateral geniculate nucleus (LGN)** of the thalamus, which acts as a relay station. From the LGN, the visual information is sent to the **primary visual cortex** in the occipital lobe at the back of the brain. Here, the brain begins to reconstruct the visual scene, processing elements like edges, colors, and motion.

The brain integrates these basic elements to form a coherent image, allowing us to recognize objects, perceive depth, and navigate our environment. The visual system is a highly complex and efficient mechanism that enables us to interpret the vast array of visual stimuli we encounter every day.

The Auditory System

The auditory system allows us to perceive sound by converting sound waves into electrical signals that the brain can interpret. This process begins when sound waves enter the **outer ear** and travel through the **ear canal** to the **eardrum** (tympanic membrane). The sound waves cause the eardrum to vibrate, initiating the hearing process.

These vibrations are then transferred to the **middle ear**, where three tiny bones—the **malleus** (hammer), **incus** (anvil), and **stapes** (stirrup)—amplify and transmit the sound to the **inner ear**. The stapes connects to the **oval window**, a membrane-covered opening that leads to the **cochlea**.

The cochlea, a fluid-filled, snail-shaped structure, is where sound vibrations are transformed into neural signals. Inside the cochlea is the **basilar membrane**, which runs along its length and is lined with thousands of **hair cells**. These hair cells are the sensory receptors of the auditory system. As the sound waves travel through the fluid in the cochlea, they cause the basilar membrane to move, which in turn bends the hair cells.

The bending of these hair cells opens ion channels, leading to the release of neurotransmitters. These neurotransmitters then stimulate the auditory nerve fibers, generating electrical signals. These signals travel along the **auditory nerve** to the **brainstem**, where initial processing occurs.

From the brainstem, the auditory signals are relayed to the **medial geniculate nucleus (MGN)** of the thalamus, and then to the **primary auditory cortex** in the temporal lobe of the brain. The primary auditory cortex is responsible for interpreting basic aspects of sound, such as pitch, volume, and rhythm.

Further processing in higher auditory areas allows us to recognize more complex sounds, such as speech and music. The brain's ability to localize sounds, or determine where they are coming from, is achieved by comparing the slight differences in the timing and intensity of sounds arriving at each ear.

The auditory system is finely tuned to detect a wide range of sounds, enabling us to communicate, enjoy music, and respond to our environment.

Somatosensory System

The somatosensory system is responsible for processing sensory information from the body, including touch, temperature, pain, and proprioception (the sense of body position). It allows us to perceive our environment and our own physical state, providing critical feedback for movement and interaction with the world.

This system begins with sensory receptors located throughout the skin, muscles, joints, and internal organs. These receptors are specialized to detect different types of stimuli. **Mechanoreceptors** respond to pressure and vibration, **thermoreceptors** detect temperature changes, **nociceptors** signal pain, and **proprioceptors** provide information about the position and movement of our body parts.

When a sensory receptor is activated, it generates an electrical signal that is transmitted via sensory nerves to the spinal cord. The signals are then relayed to the brain, where they are processed in the **somatosensory cortex** located in the parietal lobe. The somatosensory cortex is organized in a way that reflects a map of the body, known as the **somatotopic map** or homunculus, where different regions of the cortex correspond to specific body parts.

The somatosensory system is highly adaptive, allowing us to respond to changes in our environment quickly. For instance, when you touch a hot surface, nociceptors send rapid pain signals to the brain, prompting an immediate withdrawal response. Similarly, proprioceptors allow us to move smoothly and coordinate complex actions without constantly looking at our limbs.

This system's ability to integrate various types of sensory input is essential for activities such as gripping objects, feeling textures, and maintaining balance. It also plays a critical role in protective reflexes, such as pulling away from something sharp or adjusting posture to avoid injury.

Olfactory and Gustatory Systems

The olfactory system is our sense of smell, enabling us to detect and identify thousands of different odors. This system begins in the nasal cavity, where odor molecules enter and dissolve in the mucus lining the **olfactory epithelium**. The olfactory epithelium contains **olfactory receptor neurons**, which are specialized cells equipped with receptors that bind to specific odor molecules.

Each olfactory receptor neuron expresses only one type of receptor, but there are hundreds of different receptors, allowing us to detect a wide range of odors. When an odor molecule binds to its corresponding receptor, it triggers an electrical signal in the neuron. This signal is then transmitted through the axons of the olfactory receptor neurons, which converge to form the **olfactory nerve**.

The olfactory nerve carries these signals to the **olfactory bulb**, a structure located just above the nasal cavity in the brain. In the olfactory bulb, the signals are processed and refined as they pass through structures called **glomeruli**, where input from similar receptors is combined. From the olfactory bulb, the signals are sent to the **olfactory cortex** and other brain areas, including the limbic system, which is involved in emotion and memory.

The connection between the olfactory system and the limbic system explains why certain smells can evoke strong memories or emotions. Unlike other sensory systems, the olfactory system does not pass through the thalamus before reaching the cortex, making it unique in its direct connection to the brain.

This system is essential not only for detecting odors but also for recognizing food, detecting hazards (like smoke or gas), and social interactions, as many social cues in animals and humans are mediated by scent.

Gustatory System

The gustatory system is responsible for our sense of taste, allowing us to detect and differentiate between sweet, salty, sour, bitter, and umami (savory) flavors. This system is closely linked to the olfactory system, as much of what we perceive as taste is actually a combination of taste and smell.

Taste perception begins with **taste buds**, which are clusters of sensory cells located primarily on the tongue, but also on the soft palate, the inner cheeks, and the throat. Each taste bud contains taste receptor cells that are sensitive to one of the five basic tastes. When food or drink enters the mouth, chemicals dissolve in saliva and bind to receptors on the taste cells.

These interactions generate electrical signals that are transmitted via three cranial nerves—the **facial nerve (VII)**, the **glossopharyngeal nerve (IX)**, and the **vagus nerve (X)**—to the **gustatory cortex** located in the insula and frontal operculum regions of the brain. The gustatory cortex processes these signals, allowing us to perceive and identify different tastes.

In addition to identifying flavors, the gustatory system is vital in digestion by stimulating the release of saliva and digestive enzymes in response to the presence of food. It also helps detect potentially harmful substances, as bitter and sour tastes often signal spoiled or toxic foods.

Taste preferences and aversions are shaped by experience, culture, and genetics. The system is also influenced by other sensory inputs, such as texture and temperature, as well as by the olfactory system, which enhances the perception of complex flavors.

The Vestibular System

The vestibular system is essential for maintaining balance, posture, and spatial orientation. Located within the inner ear, this system detects changes in head position and motion, allowing the brain to make necessary adjustments to keep us upright and coordinated.

The vestibular system consists of two main components: the **semicircular canals** and the **otolith organs** (the utricle and saccule). The semicircular canals are three fluid-filled loops, each oriented in a different plane—horizontal, anterior, and posterior—corresponding to different directions of head movement (such as nodding, tilting, or turning). When the head moves, the fluid inside the semicircular canals lags behind due to inertia, causing the **cupula**, a gelatinous structure within the canals, to bend. This bending stimulates the **hair cells** embedded in the cupula, generating electrical signals that are sent to the brain via the vestibular nerve.

The otolith organs, the utricle and saccule, detect linear acceleration and the effects of gravity. Each of these organs contains a layer of hair cells topped with tiny calcium carbonate crystals called **otoliths**. When the head tilts or moves in a straight line, the otoliths shift, pulling on the hair cells and sending signals to the brain about the head's position relative to gravity.

These signals from the vestibular system are integrated with visual and proprioceptive inputs in the brain to maintain balance and orientation. For example, when you turn your head, the vestibular system helps stabilize your gaze by coordinating eye movements through the **vestibulo-ocular reflex (VOR)**. This reflex ensures that your eyes move in the opposite direction of your head's movement, keeping your vision stable.

Dysfunction in the vestibular system can lead to symptoms like vertigo, dizziness, and imbalance. Conditions such as **Benign Paroxysmal Positional Vertigo (BPPV)** occur when otoliths dislodge and move into the semicircular canals, causing false signals of motion. The vestibular system's critical role in maintaining equilibrium makes it indispensable for daily activities, from walking to driving.

Proprioception and Kinesthesia

Proprioception and kinesthesia are closely related senses that enable us to perceive the position and movement of our body parts without relying on vision. While they often work together, proprioception refers specifically to the sense of body position, and kinesthesia refers to the sense of body movement.

Proprioception involves the ability to detect the position of our limbs and joints in space. This sense is mediated by specialized sensory receptors located in muscles, tendons, and joints, known as **proprioceptors**. The primary types of proprioceptors are **muscle spindles**, which monitor the length and stretch of muscles, and **Golgi tendon organs**, which detect changes in muscle tension. These receptors send continuous feedback to the brain about the position and status of the body, allowing us to maintain posture, balance, and coordination.

Kinesthesia, on the other hand, is the perception of movement through space. It is closely linked to proprioception but specifically refers to the detection of motion. For instance, when you raise your arm, kinesthetic receptors in the muscles and joints provide real-time feedback to the brain about the speed and direction of the movement. This information is crucial for performing smooth and coordinated actions, such as reaching for an object or walking without tripping.

The brain integrates proprioceptive and kinesthetic information with visual and vestibular inputs to create a coherent sense of body awareness. This integration allows us to move efficiently and accurately without constantly watching our limbs. For example, you can touch your nose with your eyes closed because proprioceptors inform the brain of your hand's position relative to your face.

Damage to proprioceptive pathways, such as from neurological disorders like **peripheral neuropathy**, can severely impair balance and coordination, leading to difficulties in performing everyday tasks. Kinesthetic impairments can also affect motor learning and the execution of complex movements.

Together, proprioception and kinesthesia form the foundation of our ability to interact with the environment, enabling us to perform activities ranging from simple tasks like picking up a cup to complex movements like playing a musical instrument or engaging in sports.

CHAPTER 6: MOTOR SYSTEMS: FROM THOUGHT TO ACTION

Motor Cortex and Control of Movement

The motor cortex is the region of the brain that is vital in planning, controlling, and executing voluntary movements. It is located in the frontal lobe, just anterior to the central sulcus, and is divided into three main areas: the **primary motor cortex (M1)**, the **premotor cortex**, and the **supplementary motor area (SMA)**. Each of these areas contributes uniquely to the control of movement.

Primary Motor Cortex (M1)

The primary motor cortex is the brain's main output center for motor commands. Neurons in M1 are arranged in a somatotopic map, meaning that different parts of the cortex correspond to specific parts of the body. This map, often referred to as the **motor homunculus**, illustrates that areas controlling movements of the hands, face, and tongue occupy a larger portion of M1 due to the fine motor control required by these regions.

When you decide to move, neurons in the primary motor cortex send signals down through the **corticospinal tract**, a pathway that runs from the cortex to the spinal cord. These signals then travel to the spinal motor neurons, which directly innervate muscles, causing them to contract and produce movement. The precision and strength of these movements are finely tuned by the activity of neurons in M1.

Premotor Cortex

The premotor cortex lies just in front of the primary motor cortex and is involved in the planning and selection of movements. It plays a critical role in preparing the body for movement, particularly in response to external cues. For example, when you see a ball coming toward you and prepare to catch it, the premotor cortex is engaged in organizing the necessary movements before they are executed by M1.

The premotor cortex also helps in the coordination of complex movements involving multiple body parts, such as grasping an object while stabilizing your posture. It integrates sensory information with motor planning to ensure movements are appropriate and effective.

Supplementary Motor Area (SMA)

The supplementary motor area is involved in the initiation of movement, particularly those that are internally driven rather than triggered by external stimuli. For instance, when you decide to stand up from a chair without any specific external prompt, the SMA is actively involved in organizing this voluntary movement.

The SMA is also crucial for coordinating sequences of movements and for bilateral coordination—movements that involve both sides of the body, such as clapping or walking.

Together, these regions of the motor cortex work in concert to translate thoughts and intentions into precise and coordinated actions. Whether reaching for a cup, typing on a keyboard, or running a marathon, the motor cortex is central to the control of voluntary movement, ensuring that our actions are smooth, purposeful, and responsive to our environment.

Basal Ganglia and Movement Regulation

The basal ganglia are a group of subcortical nuclei in the brain that play a vital role in regulating movement. Located deep within the cerebral hemispheres, the basal ganglia include structures such as the **caudate nucleus**, **putamen**, **globus pallidus**, **subthalamic nucleus**, and **substantia nigra**. These nuclei work together to facilitate smooth, coordinated movements and to inhibit unwanted or excessive motions.

One of the primary functions of the basal ganglia is to regulate the initiation and termination of voluntary movements. The basal ganglia receive input from the cerebral cortex, particularly from areas involved in planning and executing movements. This input is processed through a series of direct and indirect pathways within the basal ganglia, each modulating motor activity differently.

The direct pathway facilitates movement by promoting the initiation of desired actions. When activated, it sends signals that reduce inhibitory output from the basal ganglia to the thalamus, which then sends excitatory signals back to the motor cortex, allowing the execution of movement. **The indirect pathway**, on the other hand, inhibits movements by increasing inhibitory signals to the thalamus, thereby preventing unnecessary or excessive motor activity.

The balance between these two pathways is crucial for normal motor function. Disruption in this balance can lead to movement disorders. For example, in Parkinson's disease, the degeneration of dopamine-producing neurons in the substantia nigra leads to an overactive indirect pathway and an underactive direct pathway, resulting in symptoms like bradykinesia (slowness of movement), rigidity, and tremors.

Conversely, Huntington's disease, which involves the degeneration of neurons in the striatum (caudate nucleus and putamen), leads to a decreased inhibition of the thalamus, causing excessive and involuntary movements known as chorea.

The basal ganglia are not just involved in motor control; they also play roles in cognitive and emotional functions, influencing habits, decision-making, and motivation. However, their primary contribution remains in fine-tuning motor commands to ensure that movements are smooth, purposeful, and well-regulated.

Cerebellum and Coordination

The cerebellum, located at the back of the brain beneath the occipital lobes, is essential for coordinating voluntary movements and maintaining balance and posture. Although it does not initiate movement, the cerebellum ensures that movements are smooth, precise, and coordinated.

The cerebellum receives input from multiple sources, including the motor cortex, spinal cord, and sensory systems. This input provides the cerebellum with information about the intended movement (from the motor cortex) and the current position and movement of the body (from the spinal cord and sensory systems). By comparing the intended movement with the actual movement, the cerebellum can detect any discrepancies and send corrective signals to the motor cortex and spinal cord.

One of the cerebellum's key functions is to fine-tune motor commands. When you reach out to grab a cup, for example, the motor cortex sends the initial command to the muscles involved in the movement. The cerebellum monitors the execution of this movement, adjusting the force, speed, and direction as needed to ensure that your hand reaches the cup smoothly and accurately. If the movement is off-target, the cerebellum makes rapid adjustments to correct the course.

The cerebellum is also involved in motor learning, the process by which we refine and improve our movements through practice. For instance, when learning to ride a bicycle, the cerebellum helps you adapt to the balance and coordination required, gradually improving your performance as you practice.

Damage to the cerebellum can result in **ataxia**, a condition characterized by a lack of coordination, unsteady gait, and difficulty with fine motor tasks. Individuals with cerebellar damage may have trouble with tasks that require precise movements, such as writing or buttoning a shirt, and may also experience balance problems and difficulty with tasks like walking or standing.

In addition to its role in motor control, the cerebellum has been increasingly recognized for its involvement in cognitive functions, such as attention, language, and emotional regulation. However, its primary function remains in ensuring that our movements are well-coordinated, smooth, and adapted to our goals and the environment.

Spinal Cord and Reflexes

The spinal cord is a vital part of the central nervous system, serving as a communication highway between the brain and the rest of the body. It runs from the base of the brain down through the vertebral column and is responsible for transmitting motor commands from the brain to the muscles and sensory information from the body back to the brain. In addition to these functions, the spinal cord also is important in reflexes, which are automatic, involuntary responses to specific stimuli.

Structure of the Spinal Cord

The spinal cord is a cylindrical structure made up of nerve fibers that are protected by the bony vertebrae of the spine. It is organized into segments, each corresponding to a specific region of the body. The spinal cord contains **gray matter**, which consists of neuron cell bodies and is found in the center of the cord, and **white matter**, composed of myelinated nerve fibers that surround the gray matter.

The gray matter of the spinal cord is shaped like a butterfly, with dorsal (posterior) horns that receive sensory input and ventral (anterior) horns that send motor output. Sensory neurons enter the spinal cord via the dorsal roots, while motor neurons exit through the ventral roots.

Reflexes

Reflexes are rapid, automatic responses to stimuli that do not require conscious thought. They are essential for protecting the body from harm and for maintaining posture and balance. The spinal cord is the central hub for most reflexes, allowing them to occur quickly without the need for input from the brain.

The reflex arc is the neural pathway involved in a reflex. It typically involves five components:

1. **Receptor**: A sensory receptor detects the stimulus, such as touching something hot.
2. **Sensory neuron**: The sensory neuron transmits the signal from the receptor to the spinal cord.
3. **Integration center**: In the spinal cord, the signal is processed, often involving an interneuron, which relays the signal directly to a motor neuron.

4. **Motor neuron**: The motor neuron carries the signal from the spinal cord to the muscle or gland.
5. **Effector**: The effector (a muscle or gland) responds by contracting or secreting a substance, leading to the reflex action.

A classic example of a reflex is the **patellar reflex** or "knee-jerk" reflex. When the patellar tendon is tapped, sensory receptors in the muscle send a signal through the sensory neuron to the spinal cord. The signal is processed in the spinal cord and sent back through a motor neuron, causing the quadriceps muscle to contract, which results in the lower leg kicking forward.

Reflexes are not only protective but also crucial for maintaining homeostasis. For example, reflexes help regulate heart rate, breathing, and digestion. **Withdrawal reflexes**, like pulling your hand back from a hot stove, are critical for avoiding injury, while **stretch reflexes**, such as those that prevent muscles from overstretching, help maintain muscle tone and posture.

The spinal cord's ability to mediate reflexes independently of the brain highlights its importance in both everyday functioning and emergency responses. This decentralized control allows for quicker reactions, which can be crucial in protecting the body from harm.

CHAPTER 7: LEARNING AND MEMORY: HOW WE STORE INFORMATION

Types of Memory

Memory is the brain's ability to store and retrieve information over time. It can be divided into several types, each serving different purposes and operating over different time scales.

Sensory Memory

Sensory memory is the shortest type of memory, lasting only a fraction of a second. It acts as a buffer for sensory input, allowing the brain to retain impressions of sensory information after the original stimulus has ceased. For example, when you see a flash of lightning, sensory memory briefly holds the image before it fades. Sensory memory is further divided into types like **iconic memory** for visual information and **echoic memory** for auditory information.

Short-Term Memory (STM)

Short-term memory, also known as working memory, holds information temporarily for a few seconds to a minute. This type of memory is essential for tasks like remembering a phone number long enough to dial it. STM has a limited capacity, typically holding around seven items at once, a concept known as **Miller's Law**. Information in short-term memory can be transferred to long-term memory through processes like rehearsal, where you repeat the information to keep it active.

Long-Term Memory (LTM)

Long-term memory stores information for extended periods, ranging from hours to a lifetime. Unlike short-term memory, LTM has a vast capacity. It is divided into two main categories: **explicit (declarative) memory** and **implicit (non-declarative) memory**.

- **Explicit Memory**: Explicit memory involves conscious recall of information and is further divided into **episodic memory** and **semantic memory**. Episodic memory is the ability to remember personal experiences, such as your first day of school. Semantic memory involves facts and general knowledge, like knowing that Paris is the capital of France.
- **Implicit Memory**: Implicit memory operates unconsciously and includes skills and habits. **Procedural memory**, a type of implicit memory, involves motor skills like riding a bicycle or playing the piano. These memories are

formed and retrieved without conscious awareness and often remain robust even when explicit memories are impaired.

Working Memory

Working memory is a more complex form of short-term memory. It not only temporarily stores information but also manipulates it. This is critical for tasks like problem-solving, reasoning, and comprehension. For example, when doing mental math, working memory allows you to hold and manipulate numbers and operations in your mind simultaneously.

Prospective Memory

Prospective memory involves remembering to perform a planned action or intention in the future, such as remembering to take medication at a certain time. This type of memory is essential for managing daily tasks and obligations, bridging the gap between intention and action.

Each type of memory plays a distinct role in how we process, store, and retrieve information, forming the basis of our learning and daily functioning. Understanding these different types of memory helps explain the complex ways in which the brain handles information and supports cognitive processes.

Neural Circuits of Memory

Memory formation and retrieval depend on complex neural circuits that involve multiple brain regions working together. These circuits are primarily centered in the **hippocampus, prefrontal cortex**, and other related areas, each contributing uniquely to different types of memory.

The **hippocampus,** located in the medial temporal lobe, is crucial for forming new memories, especially **episodic** and **declarative memories**. When you experience something new, the hippocampus helps encode this information by integrating sensory input and linking it with existing knowledge. The hippocampus is particularly important for spatial memory, which involves navigating and remembering locations. The neural pathways in the hippocampus, including the famous **trisynaptic circuit** involving the dentate gyrus, CA3, and CA1 regions, are critical for the consolidation of memories from short-term to long-term storage.

The **prefrontal cortex** plays a significant role in **working memory** and the organization and retrieval of memories. This region helps maintain and manipulate information over short periods, which is essential for tasks like problem-solving and decision-making. The prefrontal cortex interacts closely with the hippocampus and other brain regions to support complex memory processes, such as recalling specific details about past experiences or planning future actions based on past knowledge.

Another key structure involved in memory circuits is the **amygdala**, which is essential for emotional memories. The amygdala assigns emotional significance to memories, especially those involving fear or reward, and interacts with the hippocampus to enhance the retention of emotionally charged events.

The **basal ganglia** and **cerebellum** are involved in **procedural memory**, which relates to the acquisition of skills and habits. These areas help automate actions that become second nature over time, such as riding a bicycle or typing on a keyboard.

Together, these neural circuits work in concert to process, store, and retrieve memories, enabling us to learn from experiences, adapt our behavior, and navigate the world. Disruptions in these circuits can lead to memory impairments, as seen in conditions like Alzheimer's disease, where hippocampal atrophy leads to difficulties in forming new memories.

Mechanisms of Synaptic Plasticity

Synaptic plasticity is the ability of synapses—the connections between neurons—to strengthen or weaken over time in response to activity. This dynamic property of synapses is fundamental to learning, memory, and overall brain adaptability. Two primary mechanisms of synaptic plasticity are **long-term potentiation (LTP)** and **long-term depression (LTD)**.

Long-term potentiation (LTP) is a process that strengthens synapses based on repeated use. When a synapse is repeatedly activated, the postsynaptic neuron becomes more responsive to future stimulation. This increased responsiveness is often due to the enhancement of receptor function and an increase in the number of receptors, particularly **NMDA** and **AMPA receptors**, on the postsynaptic membrane. LTP is most prominently observed in the hippocampus, where it is considered a cellular mechanism underlying learning and memory. For example, the repeated activation of synapses during the learning process leads to LTP, making it easier for these neural circuits to activate in the future and thus strengthening the memory trace.

Long-term depression (LTD), in contrast, weakens synaptic connections. LTD occurs when synapses are less frequently used, leading to a decrease in receptor density or a reduction in neurotransmitter release. This process allows the brain to prune unnecessary connections, ensuring that only the most relevant and frequently used synapses are maintained. LTD is crucial for memory refinement and flexibility, preventing the brain from becoming overloaded with irrelevant or redundant information.

Both LTP and LTD rely on calcium signaling within neurons. **High levels of calcium influx** into the postsynaptic neuron typically trigger LTP by activating enzymes like **CaMKII**, which promotes the insertion of more AMPA receptors into the synapse. **Lower levels of calcium** can lead to LTD by activating different enzymes, such as **calcineurin**, which removes receptors from the synapse.

These mechanisms of synaptic plasticity are not only essential for learning and memory but also for adapting to changes in the environment. The brain's ability to remodel synaptic connections allows for the continuous updating and refining of knowledge, skills, and behaviors throughout life. Disruptions in synaptic plasticity are implicated in various neurological disorders, including Alzheimer's disease, schizophrenia, and depression, where the normal processes of LTP and LTD may be altered.

Disorders of Memory

Memory disorders involve disruptions in the processes of storing, retrieving, or forming memories, often leading to significant impacts on daily life. These disorders can result from various factors, including neurological diseases, brain injuries, or psychological conditions.

Alzheimer's disease is one of the most well-known memory disorders. It is a progressive neurodegenerative disorder characterized by the accumulation of amyloid plaques and tau tangles in the brain, particularly in the hippocampus and other areas critical for memory. Early symptoms include difficulty remembering recent events or conversations, with progression leading to severe memory loss, confusion, and an inability to recognize familiar people or places.

Amnesia is another significant memory disorder, where a person loses the ability to recall past memories (retrograde amnesia) or form new ones (anterograde amnesia). Retrograde amnesia often occurs due to brain trauma, such as a head injury or stroke, affecting regions like the hippocampus and temporal lobes. Anterograde amnesia, often seen in conditions like **Korsakoff syndrome** (caused by chronic alcohol abuse and thiamine deficiency), involves damage to the hippocampus, impairing the brain's ability to consolidate new memories.

Dementia is a broader category that includes multiple types of cognitive decline, including memory loss, affecting daily functioning. Besides Alzheimer's, other forms of dementia, such as **vascular dementia** and **Lewy body dementia**, also feature prominent memory impairments.

Post-traumatic stress disorder (PTSD) involves memory dysfunction, particularly with intrusive memories or flashbacks of traumatic events. In PTSD, the amygdala (involved in emotional processing) and the hippocampus (involved in memory consolidation) play critical roles, leading to the over-consolidation of traumatic memories and difficulty in distinguishing between past trauma and present reality.

Disorders of memory not only affect the ability to recall information but also impact emotional well-being, social interactions, and overall quality of life. Early diagnosis and intervention can help manage symptoms and improve outcomes for individuals with memory disorders.

Consolidation and Reconsolidation

Memory consolidation is the process by which short-term memories are transformed into long-term memories. This process involves the stabilization and integration of new information into the brain's existing knowledge network, making memories more resistant to interference or decay.

Consolidation occurs in two phases: **synaptic consolidation** and **system consolidation**. **Synaptic consolidation** happens within hours of learning, involving changes at the synaptic level, such as long-term potentiation (LTP), which strengthens the connections between neurons. These changes occur primarily in the hippocampus, where initial memory traces are formed. **System consolidation** occurs over a longer period, during which memories gradually become independent of the hippocampus and are stored in the neocortex. This shift from hippocampal dependency to cortical storage is crucial for the long-term stability of memories.

Sleep has a key role in memory consolidation. During sleep, particularly during deep stages (slow-wave sleep) and REM sleep, the brain replays and reorganizes the day's experiences, helping to consolidate memories. This is why a good night's sleep is often linked to better memory retention and learning.

Reconsolidation is a related process that occurs when a previously consolidated memory is retrieved or reactivated. When a memory is recalled, it temporarily returns to a malleable state, during which it can be modified or updated before being re-stored. Reconsolidation allows the brain to integrate new information with existing memories, adapting to new experiences or correcting errors in the original memory trace.

This process of reconsolidation has significant implications for therapy, particularly in treating conditions like PTSD. By reactivating traumatic memories in a controlled environment and introducing new, non-threatening information during reconsolidation, it may be possible to reduce the emotional impact of those memories.

Both consolidation and reconsolidation are essential for maintaining the flexibility and accuracy of our memories, allowing us to learn from experience, adapt to changes, and continuously refine our understanding of the world. Disruptions in these processes can lead to memory disorders or difficulties in retaining and recalling information accurately.

Working Memory

Working memory is the brain's system for temporarily holding and manipulating information necessary for complex cognitive tasks, such as reasoning, learning, and decision-making. Unlike short-term memory, which merely stores information for a brief period, working memory actively processes and organizes this information to solve problems or complete tasks.

Components of Working Memory

Working memory consists of several components, as proposed by the influential **Baddeley and Hitch model**:

1. **The Central Executive**: This is the control center of working memory. It directs attention, manages cognitive tasks, and coordinates the flow of information between different parts of working memory. The central executive decides what information to focus on, suppresses irrelevant information, and switches between tasks as needed.
2. **The Phonological Loop**: This component handles verbal and auditory information. It consists of two parts: the phonological store, which holds spoken words for a short time, and the articulatory rehearsal process, which allows us to repeat words or numbers to keep them in memory. For example, when you mentally repeat a phone number to remember it long enough to dial, you're using the phonological loop.
3. **The Visuospatial Sketchpad**: This subsystem deals with visual and spatial information. It allows us to visualize objects, navigate our environment, and remember where things are located. When you mentally map out a route or picture an object in your mind, you're engaging the visuospatial sketchpad.
4. **The Episodic Buffer**: This component integrates information from the phonological loop, visuospatial sketchpad, and long-term memory into a coherent sequence. It provides a space where different types of information can be combined into a single, unified representation. For example, recalling a specific event from your life involves the episodic buffer drawing on both visual imagery and verbal details stored in memory.

Role in Cognitive Function

Working memory is essential for a wide range of cognitive functions. It allows us to follow conversations, perform mental arithmetic, make decisions, and plan actions.

For example, when solving a math problem, working memory helps you keep track of the numbers, apply the operations, and remember the intermediate results.

However, working memory has limited capacity. Most people can hold about 5 to 9 items in their working memory at a time. This limitation means that tasks requiring more complex manipulation of information can become challenging, especially under cognitive load.

Working Memory and Brain Regions

The prefrontal cortex plays a critical role in working memory, particularly in managing and manipulating information. Other regions, such as the parietal cortex, are involved in storing the specific content of working memory, like visual or spatial details.

Implications of Working Memory Limitations

Limitations in working memory capacity can impact various aspects of life, including academic performance, problem-solving abilities, and even social interactions. For instance, individuals with attention-deficit/hyperactivity disorder (ADHD) often have impairments in working memory, making it difficult to stay focused and manage tasks.

Training working memory through exercises and cognitive tasks can potentially enhance its capacity and efficiency, leading to better performance in tasks that require high levels of cognitive control. However, the extent to which working memory can be improved and the long-term effects of such training remain topics of ongoing research.

In essence, working memory is the cognitive workspace where we juggle and process information, enabling us to engage in complex thought and action. Its efficiency and capacity are crucial determinants of our cognitive abilities and overall mental functioning.

CHAPTER 8: EMOTION AND THE BRAIN

The Limbic System

The limbic system is a complex set of brain structures that plays a central role in regulating emotions, memory, and behavior. It serves as the brain's emotional processing hub, influencing how we experience and express emotions and how we form memories linked to those emotions. The limbic system connects to various parts of the brain, enabling a coordinated response to emotional stimuli.

Key Components of the Limbic System

The limbic system includes several key structures, each contributing to different aspects of emotional and cognitive function:

1. **Amygdala**: The amygdala is critical for processing emotions, especially fear and anger. It detects threats in the environment and triggers the "fight or flight" response, preparing the body to respond to danger. The amygdala also plays a role in forming emotional memories, particularly those related to fear. When you recall a frightening experience, the amygdala is actively involved in bringing that memory to mind.
2. **Hippocampus**: The hippocampus is essential for forming and retrieving long-term memories, particularly episodic memories of specific events. It also helps contextualize emotional responses, linking the emotional experience with the memory of the situation. For example, if you have a joyful memory of a birthday celebration, the hippocampus helps store and recall the details of that event, while the amygdala attaches the emotional significance.
3. **Hypothalamus**: The hypothalamus regulates physiological responses to emotions, such as changes in heart rate, hormone release, and other autonomic functions. It links the nervous system to the endocrine system, controlling the release of stress hormones like cortisol through the pituitary gland. The hypothalamus is also involved in basic drives like hunger, thirst, and sexual behavior, all of which can have emotional components.
4. **Thalamus**: The thalamus acts as a relay station for sensory information, directing it to appropriate areas of the brain for further processing. In the context of the limbic system, the thalamus helps integrate sensory input with emotional responses. For instance, when you hear a loud noise, the thalamus quickly processes this sound and sends the information to the amygdala, which assesses the potential threat.
5. **Cingulate Gyrus**: The cingulate gyrus plays a role in regulating emotions and pain. It is involved in linking behavioral outcomes to motivation, which is critical for adaptive decision-making. The cingulate gyrus helps modulate

emotional responses and is particularly active when you experience conflicting emotions or need to adjust your behavior based on the emotional consequences of your actions.
6. **Prefrontal Cortex**: Although not traditionally considered part of the limbic system, the prefrontal cortex interacts closely with limbic structures to regulate emotions and decision-making. It is involved in higher-order functions like planning, impulse control, and social behavior. The prefrontal cortex can inhibit the amygdala's emotional responses, allowing for more rational decision-making in emotionally charged situations.

Function and Interactions

The limbic system's primary function is to process and regulate emotions. It determines how we react to emotional stimuli, from instinctual reactions like fear to more complex feelings like love and empathy. The interactions between the amygdala, hippocampus, and prefrontal cortex are particularly important for emotional regulation and memory.

For example, in stressful situations, the amygdala may trigger an immediate emotional response, such as fear or anxiety. However, the prefrontal cortex can assess the situation more rationally, potentially calming the amygdala's response if the threat is not as severe as initially perceived. The hippocampus might store this event as a memory, influencing how similar situations are handled in the future.

The limbic system also influences the body's physiological state in response to emotions. When faced with a stressful situation, the hypothalamus activates the autonomic nervous system, leading to increased heart rate, sweating, and the release of adrenaline. These physical changes prepare the body to deal with the emotional challenge.

Emotional memory formation is another critical function of the limbic system. Memories with strong emotional content are often more vivid and easier to recall, a process heavily influenced by the amygdala and hippocampus. This is why emotionally charged experiences, such as a traumatic event or a joyful occasion, tend to be remembered more clearly than neutral ones.

The limbic system's ability to integrate emotional, cognitive, and physiological responses makes it a central player in how we experience and interact with the world around us. Understanding its role helps explain the complexity of human emotions and behaviors, from instinctive reactions to deeply personal memories.

Emotion Regulation and Empathy

Emotion regulation is the ability to manage and respond to emotional experiences in a flexible and adaptive manner. This process involves various brain regions, with the **prefrontal cortex** (PFC) playing a central role in modulating emotional responses. Effective emotion regulation allows individuals to experience a range of emotions without becoming overwhelmed or reacting impulsively.

The PFC, particularly the **dorsolateral** and **ventromedial prefrontal cortex**, interacts with the **amygdala** to regulate emotions. The amygdala is responsible for detecting and generating emotional responses, especially to threats. When the amygdala triggers a strong emotional reaction, the PFC assesses the situation and can either dampen or sustain the response based on contextual factors. For example, if someone feels anger rising in a tense conversation, the PFC might intervene, prompting the individual to take a deep breath and consider a more measured response.

Emotion regulation strategies can be conscious, such as **cognitive reappraisal**, where a person changes their perspective on a situation to alter its emotional impact. For instance, viewing a stressful event as a challenge rather than a threat can reduce anxiety. **Suppression**, another strategy, involves inhibiting the outward expression of emotions, although this can sometimes lead to increased internal emotional stress.

Empathy, the ability to understand and share the feelings of others, is closely linked to emotion regulation. The brain regions involved in empathy include the **anterior insula**, the **anterior cingulate cortex**, and the **mirror neuron system**. The anterior insula is involved in emotional awareness, helping individuals recognize emotions in themselves and others. The anterior cingulate cortex is associated with processing social pain and empathy-related behaviors.

Empathy can be divided into two types: **affective empathy**, which is the ability to feel what another person is feeling, and **cognitive empathy**, which is the ability to understand another person's perspective or mental state. Both types of empathy are crucial for social interactions and relationships.

The ability to regulate one's own emotions can enhance empathy. When individuals can manage their emotional responses effectively, they are better able to focus on and respond to the emotions of others. This connection between emotion regulation and empathy is essential for prosocial behavior, as it allows individuals to act compassionately and thoughtfully in social situations.

The Neurochemistry of Emotion

The neurochemistry of emotion involves various neurotransmitters and hormones that influence how we experience and respond to emotions. These chemicals play key roles in mood regulation, stress responses, and the overall emotional state of an individual.

Serotonin is one of the most well-known neurotransmitters associated with mood regulation. It helps stabilize mood, feelings of well-being, and happiness. Low levels of serotonin are linked to depression, anxiety, and mood disorders. Serotonin also plays a role in regulating sleep, appetite, and even social behavior. Medications like selective serotonin reuptake inhibitors (SSRIs), commonly prescribed for depression, work by increasing serotonin levels in the brain, thereby improving mood and emotional stability.

Dopamine is another critical neurotransmitter involved in the brain's reward system. It is associated with feelings of pleasure, motivation, and reinforcement of behaviors. When you experience something enjoyable, such as eating a favorite food or receiving praise, dopamine is released, reinforcing the behavior that led to those positive feelings. Dopamine imbalances are linked to conditions like Parkinson's disease, where dopamine-producing neurons degenerate, and to addiction, where excessive dopamine activity reinforces substance use.

Norepinephrine, also known as noradrenaline, is involved in the body's "fight or flight" response. It increases arousal, alertness, and focus during stressful situations. Norepinephrine is produced in the **locus coeruleus** and released in response to stress, helping the body prepare for action. However, chronic stress can lead to an overproduction of norepinephrine, contributing to anxiety disorders and hypervigilance.

Cortisol is a hormone released by the adrenal glands in response to stress. It plays a significant role in the body's stress response by increasing blood sugar levels, enhancing the brain's use of glucose, and curbing non-essential functions during a fight or flight situation. While cortisol is crucial for managing acute stress, chronic high levels of cortisol can lead to health problems such as impaired cognitive function, anxiety, and depression.

Oxytocin is often referred to as the "love hormone" due to its role in social bonding, trust, and emotional connection. It is released in response to social interactions, such as hugging, and during childbirth and breastfeeding, enhancing the bond between mother and child. Oxytocin's effects on emotion include promoting feelings of trust and reducing fear and anxiety, making it essential for forming and maintaining social relationships.

Together, these neurochemicals work in a finely tuned balance to regulate our emotions and influence our behavior. Disruptions in this balance can lead to emotional disturbances and mental health disorders, highlighting the importance of neurochemistry in understanding and managing emotions.

Disorders of Emotion

Disorders of emotion involve disruptions in the brain's ability to regulate, process, or express emotions appropriately. These disorders can significantly impact a person's mood, behavior, and overall quality of life. They often result from a combination of genetic, environmental, and neurochemical factors.

Depression

Depression is one of the most common emotional disorders, characterized by persistent feelings of sadness, hopelessness, and a lack of interest or pleasure in activities. It is associated with imbalances in neurotransmitters such as **serotonin**, **dopamine**, and **norepinephrine**. The **prefrontal cortex** and **amygdala** are key brain regions involved in depression. The prefrontal cortex, which is responsible for regulating emotions and decision-making, often shows reduced activity in individuals with depression. The amygdala, which processes emotions, may become overactive, particularly in response to negative stimuli, leading to heightened feelings of fear or sadness.

Anxiety Disorders

Anxiety disorders, including generalized anxiety disorder (GAD), panic disorder, and social anxiety disorder, are characterized by excessive fear, worry, or nervousness. These disorders involve heightened activity in the **amygdala**, which leads to an exaggerated response to perceived threats. The **hippocampus**, which is involved in forming memories, can also play a role, particularly in disorders like PTSD, where the brain struggles to process and integrate traumatic memories. Neurotransmitters such as **norepinephrine** and **GABA** are crucial in regulating anxiety; imbalances can lead to heightened arousal and difficulty in calming the mind.

Bipolar Disorder

Bipolar disorder is characterized by extreme mood swings, including periods of depression and episodes of mania or hypomania. During manic episodes, individuals may feel excessively happy, energetic, or irritable, while depressive episodes mirror the symptoms of major depression. Bipolar disorder involves dysregulation in brain circuits that control mood, including the **prefrontal cortex**, **amygdala**, and **ventral striatum**. The neurotransmitters **dopamine** and **glutamate** are implicated in the disorder, with dopamine playing a role in the manic phase and glutamate in the regulation of mood swings.

Post-Traumatic Stress Disorder (PTSD)

PTSD develops after exposure to a traumatic event, leading to symptoms such as flashbacks, nightmares, and severe anxiety. The disorder is associated with hyperactivity in the amygdala, which becomes overly responsive to triggers associated with the trauma. The **hippocampus** may also shrink in size due to prolonged stress, affecting memory processing and the ability to distinguish between past and present. PTSD is linked to abnormalities in the regulation of cortisol and norepinephrine, which can exacerbate the body's stress response.

Borderline Personality Disorder (BPD)

BPD is characterized by intense and unstable emotions, impulsive behaviors, and difficulties in maintaining relationships. Individuals with BPD often experience rapid mood swings, chronic feelings of emptiness, and a heightened fear of abandonment. The disorder is associated with dysfunction in the limbic system, particularly the amygdala and prefrontal cortex. The amygdala's hyperactivity can lead to exaggerated emotional responses, while the prefrontal cortex may struggle to regulate these emotions effectively. Neurotransmitter imbalances involving **serotonin** and **dopamine** are also implicated in BPD, contributing to mood instability and impulsive behaviors.

Obsessive-Compulsive Disorder (OCD)

OCD is characterized by persistent, unwanted thoughts (obsessions) and repetitive behaviors (compulsions) that an individual feels compelled to perform. The disorder is linked to dysfunction in the **cortico-striato-thalamo-cortical (CSTC) circuit**, which involves the orbitofrontal cortex, the caudate nucleus, and the thalamus. This circuit is responsible for filtering out irrelevant thoughts and controlling habitual behaviors. Imbalances in **serotonin** are thought to play a significant role in OCD, and medications that enhance serotonin activity, such as SSRIs, are commonly used in treatment.

Phobias

Phobias are intense, irrational fears of specific objects, situations, or activities. The fear response in phobias is triggered by the amygdala, which becomes overly sensitive to the phobic stimulus. For example, in a person with a phobia of spiders (arachnophobia), the amygdala may react strongly even to an image of a spider, triggering a fight-or-flight response. The **insula** and the **anterior cingulate cortex** are also involved in the heightened emotional response seen in phobias.

These disorders of emotion highlight the complexity of the brain's emotional regulation systems. Each disorder involves specific brain regions and neurotransmitters, and understanding these mechanisms is crucial for developing effective treatments. Emotional disorders can be debilitating, but with appropriate intervention, individuals can manage symptoms and improve their quality of life.

CHAPTER 9: COGNITION AND EXECUTIVE FUNCTION

The Prefrontal Cortex

The prefrontal cortex (PFC) is the brain's executive control center, responsible for complex cognitive functions, decision-making, and social behavior. Located at the front of the frontal lobes, the PFC is involved in tasks that require planning, problem-solving, and regulating emotions. It is essential for what we refer to as "executive functions," which are higher-order processes that allow us to navigate complex, goal-directed activities.

Role in Executive Functions

The PFC oversees various executive functions, including working memory, cognitive flexibility, and inhibitory control. **Working memory** is the ability to hold and manipulate information in mind over short periods. For instance, when you perform mental arithmetic or follow multi-step instructions, your PFC is actively engaged in keeping track of the information and using it to guide your actions.

Cognitive flexibility refers to the capacity to adapt to new rules or switch between tasks. The PFC allows you to adjust your behavior when the environment changes, such as shifting your strategy when solving a problem or responding to a change in plans. This flexibility is critical for effective decision-making and adapting to unexpected situations.

Inhibitory control is the ability to suppress impulses and distractions, enabling you to focus on the task at hand. This function is essential for self-regulation, allowing you to resist temptations, avoid rash decisions, and stay on course toward long-term goals.

Decision-Making and Social Behavior

The PFC plays a central role in decision-making, particularly in weighing the potential consequences of actions. It integrates information from other brain regions, such as the limbic system (which processes emotions) and the parietal cortex (which handles sensory information), to make informed choices. The PFC helps balance emotional impulses with rational thought, ensuring that decisions are both practical and aligned with long-term objectives.

In social contexts, the PFC is involved in understanding social norms, empathy, and moral reasoning. It enables you to consider the perspectives of others, predict how they might react, and adjust your behavior accordingly. This social cognition is crucial for maintaining relationships and functioning effectively in society.

Development and Plasticity

The PFC is one of the last brain regions to fully mature, continuing to develop into early adulthood. This extended development period is linked to the gradual acquisition of executive functions and the ability to make complex decisions. The PFC's plasticity, or ability to adapt based on experience, allows for the refinement of these functions throughout life. For example, repeated practice in decision-making or problem-solving can strengthen the neural connections within the PFC, enhancing cognitive abilities.

Dysfunction and Disorders

Dysfunction in the PFC is associated with a range of cognitive and behavioral disorders. **Attention-deficit/hyperactivity disorder (ADHD)**, for example, involves impairments in inhibitory control and working memory, often linked to reduced activity in the PFC. Similarly, damage to the PFC, whether from injury or neurodegenerative diseases, can lead to difficulties in planning, social behavior, and impulse control.

The prefrontal cortex is a critical hub for the cognitive processes that define human intelligence and behavior. Its ability to integrate information, regulate emotions, and guide decision-making makes it essential for navigating the complexities of everyday life. Understanding the PFC's functions and how they contribute to cognition provides insights into how we think, plan, and interact with the world around us.

Attention and Perception

Attention and perception are closely linked cognitive processes that allow us to navigate and make sense of the world around us. **Attention** is the brain's ability to focus on specific stimuli or information while filtering out distractions. **Perception** is the process of interpreting sensory information to create a meaningful experience of the environment. Together, they enable us to process and respond to the vast amount of information we encounter every moment.

Selective attention is the ability to focus on a particular stimulus while ignoring others. For example, in a noisy room, you can concentrate on a conversation with one person while tuning out the background noise. This ability is crucial for effective perception because it allows the brain to prioritize relevant information and avoid being overwhelmed by sensory input. The **prefrontal cortex** and **parietal lobes** play key roles in controlling attention, directing cognitive resources to where they are needed most.

Sustained attention refers to the ability to maintain focus over time, which is essential for tasks that require prolonged concentration, such as reading or driving.

The brain's ability to sustain attention depends on networks that involve the prefrontal cortex, the parietal cortex, and subcortical structures like the **thalamus** and **basal ganglia**.

Perception involves interpreting sensory input from the environment to form a coherent representation of reality. This process is not just a passive reception of stimuli but involves active construction by the brain. For instance, when you look at a tree, your brain integrates visual information about the tree's shape, color, and texture, along with prior knowledge about what a tree is, to create a complete image in your mind.

Bottom-up processing and **top-down processing** are two key mechanisms in perception. **Bottom-up processing** starts with sensory input—what your senses detect—and builds up to a perception. This is driven by the actual data from the environment. In contrast, **top-down processing** involves using prior knowledge, experiences, and expectations to interpret sensory information. For example, if you expect to see a specific object in a certain location, your brain might quickly recognize it based on that expectation, even if the sensory input is incomplete.

Attention also has a key role in perception by determining what sensory information is processed further and what is discarded. **Inattentional blindness** is a phenomenon where people fail to perceive an unexpected stimulus in plain sight because their attention is focused elsewhere. This highlights how selective attention influences what we perceive, often at the expense of missing other potentially important information.

Attention and perception are fundamental to how we interact with the world. They allow us to focus on what matters, make sense of complex environments, and respond appropriately to the changing stimuli around us.

Language and the Brain

Language is one of the most complex cognitive functions, and its processing involves multiple areas of the brain working in concert. The brain's ability to produce and understand language is primarily centered in the left hemisphere for most people, with key regions including **Broca's area** and **Wernicke's area**.

Broca's area, located in the frontal lobe, is crucial for language production. It controls the motor functions involved in speech, such as the movement of the lips, tongue, and vocal cords. Damage to Broca's area can result in **Broca's aphasia**, a condition characterized by slow, halting speech and difficulty forming complete sentences, though comprehension typically remains intact. Individuals with Broca's aphasia know what they want to say but struggle to express it verbally.

Wernicke's area, found in the temporal lobe, is essential for language comprehension. It allows us to understand spoken and written language by processing the meaning of words and sentences. Damage to Wernicke's area leads to **Wernicke's aphasia**, where individuals can produce fluent speech, but it often lacks meaning or is filled with nonsensical words. They may also have difficulty understanding language, leading to significant communication challenges.

These two areas are connected by a bundle of nerve fibers known as the **arcuate fasciculus**. This connection enables the coordination between understanding language (Wernicke's area) and producing it (Broca's area). Disruption of this pathway can lead to **conduction aphasia**, where a person can understand language and speak it, but may struggle with repeating phrases or sentences.

Language processing also involves other brain regions. The **angular gyrus** plays a role in reading and writing by integrating visual and auditory information, while the **supramarginal gyrus** is involved in phonological processing, helping to decode spoken words. The right hemisphere contributes to the emotional tone and prosody of speech, adding nuance and emphasis to communication.

The brain's language network is also highly adaptable. **Neuroplasticity** allows other regions to compensate for damaged areas, especially in children, whose brains are more flexible. For instance, if Broca's area is damaged early in life, nearby regions or even the corresponding area in the right hemisphere might take over some of its functions, allowing for a degree of language recovery.

Bilingualism adds another layer of complexity to the brain's language processing. Bilingual individuals often have enhanced connectivity in the brain regions involved in executive function, as switching between languages requires significant cognitive control. This increased connectivity is thought to contribute to the cognitive benefits often observed in bilinguals, such as improved attention and problem-solving skills.

Language is not just about communication; it shapes how we think, perceive the world, and interact with others. The brain's ability to process language is a testament to the complexity and flexibility of our cognitive systems, allowing us to convey ideas, emotions, and knowledge across generations.

Cognitive Aging

Cognitive aging refers to the changes in cognitive abilities that occur as people grow older. While aging is associated with certain declines in cognitive function, not all cognitive abilities deteriorate at the same rate, and some may even remain stable or improve with age.

One of the most commonly observed changes is a decline in **processing speed**. This refers to the time it takes to perceive, understand, and respond to information. Older adults often experience slower reaction times, which can affect tasks that require quick decision-making or rapid responses. This slowing is partly due to changes in the brain's white matter, which is involved in the transmission of signals between different brain regions.

Working memory—the ability to hold and manipulate information over short periods—also tends to decline with age. This can make it more challenging for older adults to multitask or manage complex cognitive tasks. The **prefrontal cortex**, which plays a key role in working memory, often shows signs of atrophy (shrinkage) with aging, contributing to these difficulties.

Episodic memory, or the ability to recall specific events from one's past, also tends to decline with age. The hippocampus, a region of the brain crucial for memory formation, often experiences structural changes, such as reduced volume, which can impair memory function. This is why older adults may have trouble remembering recent events but may recall long-ago memories with clarity.

However, not all cognitive functions decline with age. **Semantic memory**, which involves knowledge of facts and general information, often remains stable or can even improve as people accumulate more knowledge over their lifetime. **Vocabulary** and **language skills** also tend to remain intact, and older adults often excel in tasks that rely on accumulated knowledge and experience.

Crystallized intelligence, which involves using learned knowledge and experience, typically remains stable or improves with age. This contrasts with **fluid intelligence**, which involves reasoning and problem-solving in novel situations and tends to decline.

Cognitive reserve refers to the brain's ability to adapt to age-related changes by using alternative strategies or brain networks. Engaging in mentally stimulating activities, maintaining social connections, and leading a healthy lifestyle can help build cognitive reserve, potentially delaying or mitigating cognitive decline.

While cognitive aging is a natural process, it varies widely among individuals. Factors such as genetics, education, lifestyle, and overall health play significant roles in determining how cognitive abilities change with age. Understanding these factors can help in developing strategies to maintain cognitive health and promote successful aging.

Creativity and the Brain

Creativity is the ability to generate novel ideas, solutions, or expressions. It involves a complex interplay of cognitive processes and brain regions, allowing individuals to think outside the box, make connections between seemingly unrelated concepts, and bring new ideas to life.

Creativity engages both hemispheres of the brain, though certain regions are particularly important. The **prefrontal cortex** is important in creative thinking by supporting processes such as planning, decision-making, and cognitive flexibility. This region enables individuals to generate ideas, evaluate their feasibility, and refine them into workable solutions. **Cognitive flexibility**, or the ability to shift perspectives and approach problems from different angles, is essential for creativity and is heavily dependent on the prefrontal cortex.

Another key region involved in creativity is the **default mode network (DMN)**, a network of interconnected brain regions that is active when the mind is at rest and engaged in introspective activities, such as daydreaming or mind-wandering. The DMN includes the **medial prefrontal cortex, posterior cingulate cortex**, and **inferior parietal lobule**. When individuals are not focused on a specific task, the DMN allows for the free flow of thoughts and associations, facilitating the generation of creative ideas.

Creativity also relies on the **executive control network**, which includes regions such as the dorsolateral prefrontal cortex and anterior cingulate cortex. This network helps manage attention, inhibit irrelevant information, and maintain focus on developing ideas, ensuring that creative thoughts are organized and directed toward a goal.

The **brain's ability to make novel connections** between disparate pieces of information is a hallmark of creativity. This involves associative thinking, where the brain links ideas that are not typically related. The **temporal lobes**, particularly the anterior temporal lobe, are involved in this process, helping to retrieve and recombine information from memory to form new concepts.

Dopamine, a neurotransmitter associated with reward and motivation, also plays a role in creativity. Higher dopamine levels can enhance cognitive flexibility and the ability to see multiple possibilities, which are key components of creative thinking. This is why people often feel more creative when they are in a positive mood or when they are motivated by a rewarding task.

Creativity is not confined to artistic endeavors; it is also crucial in scientific innovation, problem-solving, and everyday decision-making. The brain's creative processes allow individuals to adapt to new challenges, find innovative solutions, and express themselves in unique ways. Understanding the neural underpinnings of creativity can help in fostering environments that encourage creative thinking and innovation across various domains.

Metacognition

Metacognition refers to the awareness and understanding of one's own thought processes. It involves thinking about thinking, allowing individuals to monitor, regulate, and direct their cognitive activities effectively. This capability is important in learning, problem-solving, and decision-making, as it enables people to assess their knowledge, strategies, and cognitive limitations.

Components of Metacognition

Metacognition is typically divided into two main components: **metacognitive knowledge** and **metacognitive regulation**.

1. **Metacognitive Knowledge**: This refers to what individuals know about their cognitive processes. It includes understanding one's strengths and weaknesses, knowledge of strategies for learning and problem-solving, and awareness of the tasks at hand. For example, a student who knows they learn best by visualizing information is using metacognitive knowledge to choose study strategies that work for them.
    - **Declarative knowledge** involves knowing what strategies are available and what factors influence learning and memory.
    - **Procedural knowledge** involves knowing how to use these strategies.
    - **Conditional knowledge** involves knowing when and why to use certain strategies based on the situation.
2. **Metacognitive Regulation**: This involves the control and regulation of cognitive processes through planning, monitoring, and evaluating. Metacognitive regulation helps individuals guide their learning and problem-solving activities, adjust strategies as needed, and evaluate the effectiveness of their cognitive efforts.
    - **Planning** involves selecting appropriate strategies and allocating resources before undertaking a task. For instance, before starting a complex assignment, a person might decide to break it down into smaller, manageable parts.
    - **Monitoring** refers to the ongoing assessment of one's performance during a task. This might involve checking if a reading comprehension strategy is working or if a problem-solving approach needs adjustment.
    - **Evaluating** involves reviewing and reflecting on the outcome of a task, assessing the effectiveness of the strategies used, and making adjustments for future tasks.

The Role of Metacognition in Learning

Metacognition is essential for effective learning. Students who use metacognitive strategies are often better at understanding new material, adapting to different types of tasks, and overcoming challenges. For example, during an exam, a student with strong metacognitive skills can recognize when they do not understand a question, pause, and employ strategies to decode it, rather than just guessing.

Metacognition also encourages **self-regulated learning**, where individuals take control of their own learning process. This includes setting goals, selecting appropriate study methods, and adjusting approaches based on ongoing feedback. Self-regulated learners tend to be more successful because they can adapt to different learning environments and challenges.

Metacognition in Problem-Solving and Decision-Making

In problem-solving, metacognition allows individuals to evaluate the effectiveness of different strategies and choose the best approach. For example, when faced with a complex problem, a person might reflect on past experiences, consider various solutions, and monitor their progress, making adjustments as needed.

In decision-making, metacognition involves recognizing when a decision is complex, identifying potential biases, and assessing the reliability of the information available. This reflective process helps in making more informed and balanced decisions, particularly in situations where quick judgments might lead to errors.

Developing Metacognitive Skills

Metacognitive skills can be developed and improved through practice and reflection. Educators can foster metacognition in students by encouraging them to think about their thinking, use reflective questioning, and engage in activities that promote planning, monitoring, and evaluating their learning processes.

In everyday life, individuals can enhance their metacognitive abilities by regularly reflecting on their thought processes, being mindful of cognitive biases, and experimenting with different strategies for learning and problem-solving. By becoming more metacognitively aware, people can improve their ability to learn, make decisions, and solve problems effectively.

CHAPTER 10: SLEEP AND CIRCADIAN RHYTHMS

Stages of Sleep

Sleep is a complex process that has an important role in physical and mental health. It consists of several stages, each with distinct characteristics and functions. Understanding these stages provides insight into how the brain and body restore themselves during sleep.

Stage 1: NREM Sleep (Non-Rapid Eye Movement)

Stage 1 is the lightest stage of sleep, serving as a transition between wakefulness and sleep. During this stage, the brain produces **theta waves**, which are slower than the brainwaves seen during wakefulness. You might experience **hypnic jerks** or sudden muscle contractions, which can sometimes jolt you awake. This stage usually lasts only a few minutes, as the body begins to relax, and the heart rate slows down. It's easy to wake up from this stage, and if you do, you might not even realize you had fallen asleep.

Stage 2: NREM Sleep

Stage 2 marks the onset of true sleep. During this stage, the brain's activity slows further, characterized by **sleep spindles** (brief bursts of rapid brain activity) and **K-complexes** (sharp waves that are believed to play a role in memory consolidation). The body temperature drops, and the heart rate continues to slow. Stage 2 sleep is important for maintaining a state of restful sleep and accounts for about 50% of total sleep time. Although still considered light sleep, it is slightly more challenging to wake up from than Stage 1.

Stage 3: NREM Sleep (Deep Sleep)

Stage 3 is also known as **slow-wave sleep (SWS)** or deep sleep. During this stage, the brain produces **delta waves**, which are the slowest and highest amplitude brainwaves. This stage is crucial for physical restoration, as the body repairs tissues, builds bone and muscle, and strengthens the immune system. Deep sleep is also important for cognitive functions like memory consolidation and learning. Waking up during this stage can leave you feeling groggy and disoriented, a phenomenon known as **sleep inertia**.

Stage 4: REM Sleep (Rapid Eye Movement)

REM sleep is the stage associated with vivid dreaming. Unlike the previous stages, brain activity during REM sleep is similar to that during wakefulness, with **beta waves** and **alpha waves** dominating. The eyes move rapidly beneath the closed eyelids, and the body experiences **atonia**, a temporary paralysis of most muscles, preventing you from acting out your dreams. REM sleep is vital for emotional regulation, creativity, and memory processing, particularly for consolidating procedural and emotional memories. REM sleep occurs in cycles throughout the night, with each cycle lasting longer than the previous one, eventually making up about 20-25% of total sleep time.

Sleep Cycles

The stages of sleep repeat in cycles throughout the night, typically lasting about 90 minutes each. A full night's sleep includes multiple cycles, with the proportion of REM sleep increasing and deep sleep decreasing as the night progresses. This cyclical nature of sleep is essential for achieving the full restorative benefits that sleep provides.

Understanding the stages of sleep highlights the importance of a full night's rest, as each stage contributes to different aspects of health and well-being. Disruptions to any stage can impact physical restoration, cognitive function, and emotional health.

Neural Mechanisms of Sleep Regulation

The regulation of sleep is controlled by a complex interaction of neural circuits and chemicals within the brain. Two primary systems govern sleep: the **homeostatic sleep drive** and the **circadian rhythm**. Together, these systems determine when we feel sleepy and when we are alert.

The **homeostatic sleep drive** operates on the principle that the longer you are awake, the more you need sleep. This drive is influenced by the accumulation of a molecule called **adenosine** in the brain. As you stay awake, adenosine levels increase, creating pressure to sleep. Adenosine inhibits neurons in the **wake-promoting regions** of the brain, particularly in the **basal forebrain** and **hypothalamus**, making you feel drowsy. Sleep reduces adenosine levels, resetting the drive and restoring alertness.

The **circadian rhythm** is a roughly 24-hour cycle that regulates sleep and wakefulness, among other bodily functions. The **suprachiasmatic nucleus (SCN)**, located in the hypothalamus, is the master clock that coordinates this rhythm. The SCN receives direct input from the eyes, particularly from specialized retinal ganglion cells that detect light. This light input helps the SCN synchronize the circadian rhythm with the external environment, promoting wakefulness during the day and sleep at night.

Within the brainstem, the **reticular activating system (RAS)** is a critical network that maintains wakefulness by activating the cortex. It releases neurotransmitters like **norepinephrine** from the **locus coeruleus** and **serotonin** from the **raphe nuclei**, both of which promote alertness. Additionally, the **orexin (hypocretin)**

neurons in the hypothalamus are important in stabilizing wakefulness and preventing sudden transitions into sleep. Deficiency in orexin can lead to narcolepsy, a condition characterized by excessive daytime sleepiness and sudden sleep attacks.

On the flip side, the **ventrolateral preoptic nucleus (VLPO)** in the hypothalamus is essential for initiating sleep. The VLPO inhibits wake-promoting regions by releasing **GABA** (gamma-aminobutyric acid) and **galanin**, which dampen the activity of the arousal systems. This inhibition allows the brain to transition into sleep, particularly into non-REM sleep.

REM sleep is regulated by a different set of mechanisms involving the **pons** and **medulla**. During REM, specific neurons in these areas activate and produce **acetylcholine**, which promotes the brain activity seen in this stage, while simultaneously inhibiting motor neurons to induce atonia, or muscle paralysis, preventing the body from acting out dreams.

These neural mechanisms are finely tuned to ensure the proper balance between wakefulness and sleep. Disruptions to these systems, such as through stress, illness, or irregular sleep schedules, can lead to sleep disorders and negatively impact overall health.

Circadian Rhythms

Circadian rhythms are biological processes that follow a roughly 24-hour cycle, governing various physiological functions, including the sleep-wake cycle, hormone release, body temperature, and metabolism. These rhythms are intrinsic to the body but are influenced by external cues, primarily light, which help synchronize them with the environment.

The master regulator of circadian rhythms is the **suprachiasmatic nucleus (SCN)**, a small cluster of neurons located in the hypothalamus. The SCN receives direct input from the eyes through the **retinohypothalamic tract**, which allows it to detect changes in light and darkness. Light is the most potent external cue, or **zeitgeber**, for resetting the circadian clock, helping to align the body's internal rhythms with the day-night cycle.

Within the SCN, a series of **clock genes** operates in a feedback loop to generate rhythmic activity. These genes include **CLOCK** and **BMAL1**, which promote the expression of other genes, such as **PER** and **CRY**. As PER and CRY proteins accumulate, they inhibit the activity of CLOCK and BMAL1, creating a cycle of activation and inhibition that lasts approximately 24 hours. This molecular clock drives the rhythmic expression of various physiological processes, including the timing of sleep and wakefulness.

Circadian rhythms regulate the release of hormones such as **melatonin** and **cortisol**. Melatonin, produced by the pineal gland, is released in response to darkness and promotes sleep by signaling to the body that it is time to rest. Melatonin levels typically rise in the evening, peak during the night, and decrease in the morning as light suppresses its production. Cortisol, known as the stress hormone, follows a circadian pattern that peaks in the early morning, helping to promote alertness and prepare the body for the day.

Circadian rhythms also influence body temperature, which typically drops during the night to facilitate sleep and rises in the morning to promote wakefulness. Metabolism, including the timing of hunger and digestion, is also under circadian control, which is why irregular eating patterns can disrupt metabolic health.

Disruptions to circadian rhythms, such as those caused by shift work, jet lag, or exposure to artificial light at night, can lead to **circadian rhythm disorders**. These disruptions can have significant health consequences, including sleep disorders, mood disturbances, and increased risk of chronic conditions like obesity, diabetes, and cardiovascular disease.

Understanding and maintaining healthy circadian rhythms is crucial for overall well-being. Strategies such as maintaining a consistent sleep schedule, getting exposure to natural light during the day, and reducing light exposure at night can help keep circadian rhythms aligned and support better health.

Sleep Disorders

Sleep disorders are conditions that affect the quality, timing, and duration of sleep, leading to daytime distress and impaired functioning. They can have a significant impact on physical health, mental well-being, and overall quality of life.

Insomnia is one of the most common sleep disorders, characterized by difficulty falling asleep, staying asleep, or waking up too early and not being able to go back to sleep. It can be acute (short-term) or chronic (lasting for three months or longer). Insomnia is often linked to stress, anxiety, depression, or poor sleep habits, but it can also be a symptom of other medical conditions. People with insomnia often experience fatigue, mood disturbances, and difficulty concentrating during the day.

Sleep apnea is another prevalent sleep disorder, marked by repeated interruptions in breathing during sleep. The most common type, **obstructive sleep apnea (OSA)**, occurs when the muscles in the throat relax too much, blocking the airway. This leads to brief awakenings throughout the night, often without the person being aware of them, resulting in fragmented sleep. Symptoms include loud snoring, choking or gasping during sleep, and excessive daytime sleepiness.

Untreated sleep apnea can lead to serious health problems, including hypertension, heart disease, and stroke.

Restless legs syndrome (RLS) is a neurological disorder characterized by an uncontrollable urge to move the legs, usually due to uncomfortable sensations. These sensations typically worsen at night, making it difficult to fall asleep and stay asleep. The exact cause of RLS is unknown, but it is thought to involve dysfunction in the brain's dopamine pathways. RLS can lead to significant sleep disturbances and is often associated with periodic limb movement disorder (PLMD), where repetitive limb movements occur during sleep.

Narcolepsy is a chronic sleep disorder characterized by excessive daytime sleepiness and sudden attacks of sleep. People with narcolepsy may fall asleep unexpectedly, even during activities like driving or eating. Narcolepsy is often accompanied by **cataplexy**, a sudden loss of muscle tone triggered by strong emotions, as well as sleep paralysis and hallucinations during the transition between sleep and wakefulness. Narcolepsy is linked to a deficiency of **orexin (hypocretin)**, a neurotransmitter that regulates wakefulness.

Circadian rhythm disorders occur when the internal body clock is misaligned with the external environment. Examples include **delayed sleep-wake phase disorder**, where individuals have difficulty falling asleep and waking up at conventional times, and **shift work disorder**, which affects people who work non-traditional hours. These disorders can lead to chronic sleep deprivation, increasing the risk of mood disorders, cognitive impairment, and other health issues.

Treating sleep disorders typically involves a combination of lifestyle changes, behavioral therapies, and, in some cases, medication. Addressing underlying conditions, maintaining good sleep hygiene, and seeking professional help when needed are essential steps in managing sleep disorders and improving overall sleep quality.

Sleep Across the Lifespan

Sleep needs and patterns change significantly across the lifespan, reflecting the different physiological and developmental requirements at each stage of life.

Infants require the most sleep of any age group, typically sleeping 14 to 17 hours a day. Newborns do not have a fully developed circadian rhythm, leading to irregular sleep patterns with frequent awakenings for feeding. As they grow, their sleep consolidates into longer periods, with a more pronounced day-night cycle emerging by three to six months of age. REM sleep occupies a significant portion of infants' sleep, playing a critical role in brain development.

Toddlers and preschoolers need about 11 to 14 hours of sleep, including naps. As children grow, the proportion of REM sleep decreases, and the time spent in deep, slow-wave sleep (Stage 3 NREM) increases, which is important for physical growth and development. Establishing consistent sleep routines during this stage is crucial for promoting healthy sleep patterns and supporting cognitive and emotional development.

School-aged children typically require 9 to 11 hours of sleep per night. During this stage, sleep is critical for learning, memory consolidation, and overall growth. Children are increasingly exposed to social and academic pressures, which can sometimes lead to sleep difficulties. Ensuring a regular sleep schedule and minimizing distractions before bedtime, such as screen time, are important for maintaining good sleep hygiene.

Adolescents often experience a shift in their circadian rhythm, leading to a natural tendency to stay up later and sleep in longer. Despite needing about 8 to 10 hours of sleep, many teens suffer from sleep deprivation due to early school start times, social activities, and increased academic demands. This sleep deprivation can impact mood, cognitive performance, and overall health, making it crucial to educate teens about the importance of sleep and encourage healthier sleep habits.

Adults generally require 7 to 9 hours of sleep per night. During adulthood, sleep quality can be affected by various factors, including stress, work demands, and lifestyle choices. Many adults experience disruptions in their sleep due to insomnia or other sleep disorders, highlighting the importance of prioritizing sleep and managing stress to maintain overall health and well-being.

Older adults may experience changes in sleep architecture, including a reduction in deep sleep and more frequent awakenings during the night. While the need for sleep does not decrease significantly with age, older adults may find it more difficult to stay asleep or sleep deeply. Sleep disorders like sleep apnea and restless legs syndrome become more common with aging, further impacting sleep quality. Maintaining a consistent sleep routine, staying physically active, and addressing medical conditions that may affect sleep are essential for promoting healthy sleep in older adults.

Throughout the lifespan, sleep remains a fundamental pillar of health, supporting physical growth, cognitive function, and emotional well-being. Understanding the changing sleep needs at each stage of life can help individuals and caregivers foster healthy sleep habits and address sleep challenges as they arise.

Dreaming and Brain Activity

Dreaming is a fascinating aspect of sleep, characterized by vivid sensory experiences and often complex narratives that occur primarily during **REM (Rapid Eye Movement) sleep**. Although dreams can happen during other sleep stages, REM sleep is particularly associated with the most vivid and memorable dreams. The study of dreaming provides insights into the intricate relationship between brain activity and consciousness.

Brain Activity During REM Sleep

During REM sleep, the brain is highly active, with electrical activity that closely resembles the awake state. This activity is driven by the **pons**, a part of the brainstem, which sends signals to the **thalamus** and subsequently to the **cerebral cortex**, the brain's outer layer responsible for higher cognitive functions. The **pons** also plays a role in inducing the atonia, or muscle paralysis, that prevents the body from acting out dreams.

The **cerebral cortex** is particularly active during REM sleep, especially in regions involved in visual processing, emotion, and memory. The **occipital lobe**, which processes visual information, contributes to the vivid imagery often experienced in dreams. The **limbic system**, including the **amygdala** and **hippocampus**, is also highly active during REM sleep, influencing the emotional content of dreams and linking them to memories.

Interestingly, the **prefrontal cortex**, which is responsible for logical reasoning and impulse control, shows reduced activity during REM sleep. This reduction may explain why dreams often lack the structure and logic seen in wakeful thinking and why they can be surreal or irrational. The diminished prefrontal activity also contributes to the suspension of critical judgment, allowing the dreamer to accept bizarre scenarios as plausible within the context of the dream.

Theories of Dreaming

Several theories attempt to explain the purpose and function of dreaming. One prominent theory is the **activation-synthesis model**, proposed by neuroscientists Allan Hobson and Robert McCarley. According to this model, dreams result from the brain's attempt to make sense of random neural activity generated during REM sleep. The cerebral cortex synthesizes this activity into a coherent narrative, even if the content is bizarre or fragmented.

Another theory is that dreaming plays a role in **memory consolidation**. During sleep, the brain processes and integrates experiences from the day, and dreaming may help solidify these memories by reactivating neural circuits involved in learning and memory. This could explain why dreams often incorporate elements from recent experiences or unresolved emotional issues.

Emotional regulation is another function attributed to dreaming. The activation of the limbic system during REM sleep allows the brain to process and work through emotional experiences, potentially helping to reduce stress and anxiety. This idea is supported by studies showing that people who experience disturbing dreams or nightmares often have unresolved emotional conflicts.

Lucid Dreaming and Brain Activity

Lucid dreaming, a phenomenon where the dreamer becomes aware they are dreaming and can sometimes control the dream, involves a unique pattern of brain activity. In lucid dreams, the **prefrontal cortex** becomes more active, allowing for greater self-awareness and cognitive control within the dream. This contrasts with typical REM sleep, where the prefrontal cortex is less active.

Lucid dreaming offers a rare glimpse into the brain's ability to maintain consciousness and cognitive functions in a dream state, blurring the line between wakefulness and sleep.

Dreaming and Neurotransmitters

Neurotransmitters also have an important role in regulating brain activity during sleep and dreaming. **Acetylcholine** is particularly important for REM sleep, promoting the transition into this stage and maintaining the heightened brain activity characteristic of dreaming. In contrast, **serotonin** and **norepinephrine** levels are lower during REM sleep, which may contribute to the distinct mental state experienced during dreaming.

Clinical Implications

Understanding the neural mechanisms of dreaming has clinical implications, particularly in the context of sleep disorders and mental health. Conditions like **REM sleep behavior disorder (RBD)**, where the muscle paralysis normally associated with REM sleep is absent, leading to physical enactment of dreams, highlight the importance of proper neural regulation during sleep. Additionally, recurrent nightmares and disturbed dreaming patterns are common in conditions like PTSD, suggesting that interventions targeting sleep and dreaming could be beneficial in treatment.

Dreaming is a complex interplay of brain activity that reflects the brain's efforts to process, integrate, and make sense of internal and external experiences. The study of dreaming not only enhances our understanding of sleep but also provides insights into the workings of the conscious mind.

CHAPTER 11: NEUROSCIENCE OF PAIN

Types of Pain

Pain is a complex sensory and emotional experience that serves as a warning system, alerting the body to potential harm. It is classified into different types based on its source, duration, and underlying mechanisms. Understanding these types of pain helps in diagnosing conditions and tailoring appropriate treatments.

Nociceptive Pain

Nociceptive pain is the most common type of pain, arising from tissue damage or inflammation. It occurs when specialized sensory neurons called **nociceptors** detect harmful stimuli, such as extreme temperatures, mechanical injury, or chemical irritants. These nociceptors are located throughout the body, including the skin, muscles, joints, and internal organs.

When activated, nociceptors send signals through peripheral nerves to the spinal cord, where they are relayed to the brain. The brain processes these signals and perceives them as pain, prompting a protective response, such as withdrawing from the source of harm. Nociceptive pain can be further categorized into **somatic pain** and **visceral pain**.

Somatic pain originates from the skin, muscles, and joints. It is typically well-localized and sharp, making it easy to identify the source. For example, a cut on the skin or a sprained ankle causes somatic pain. **Visceral pain**, on the other hand, comes from the internal organs. It is often described as a deep, squeezing, or cramping sensation and is usually harder to pinpoint. Conditions like appendicitis or irritable bowel syndrome (IBS) cause visceral pain.

Neuropathic Pain

Neuropathic pain arises from damage to or dysfunction of the nervous system itself. Unlike nociceptive pain, which is a response to actual or potential tissue damage, neuropathic pain is caused by abnormal processing of pain signals in the peripheral or central nervous system. This type of pain is often described as burning, shooting, or tingling and can be persistent and difficult to treat.

Peripheral neuropathic pain occurs when there is damage to the peripheral nerves, such as in **diabetic neuropathy** or **postherpetic neuralgia** (pain following shingles). **Central neuropathic pain** arises from damage to the brain or spinal cord, as seen in conditions like **multiple sclerosis** or after a **stroke**.

Neuropathic pain can occur even in the absence of an obvious injury, and it often responds poorly to standard pain medications like NSAIDs or opioids. Treatment usually involves medications that target nerve function, such as anticonvulsants or antidepressants.

Inflammatory Pain

Inflammatory pain is a subtype of nociceptive pain, but it is worth discussing separately due to its distinct characteristics. This type of pain results from the body's immune response to injury or infection. When tissues are damaged, immune cells release inflammatory mediators like **prostaglandins** and **cytokines**. These substances sensitize nociceptors, making them more responsive to pain signals.

Inflammatory pain is often associated with conditions like arthritis, where ongoing inflammation in the joints causes chronic pain and stiffness. Unlike other types of pain, inflammatory pain tends to be more responsive to anti-inflammatory medications, such as NSAIDs, which reduce the production of inflammatory mediators.

Functional Pain

Functional pain is a type of pain that arises from abnormal functioning of the nervous system without any obvious structural damage or inflammation. Conditions like **fibromyalgia** and **irritable bowel syndrome (IBS)** are examples of functional pain disorders. The pain in these conditions is real and often chronic, but its underlying mechanisms are not fully understood.

Functional pain is challenging to treat because it does not respond well to conventional painkillers. Management often involves a multidisciplinary approach, including medication, physical therapy, and psychological support.

Acute vs. Chronic Pain

Pain can also be classified based on its duration. **Acute pain** is short-term, usually lasting less than three months, and serves as a warning signal for injury or illness. It typically resolves once the underlying cause is treated. **Chronic pain**, however, persists for more than three months and may continue even after the initial injury has healed. Chronic pain can be debilitating, leading to significant physical and emotional distress.

Understanding the different types of pain is crucial for effective pain management. Each type involves distinct mechanisms and requires tailored approaches for treatment, highlighting the complexity of the brain's role in processing and interpreting pain signals.

Pain Pathways in the Brain

Pain perception begins with the activation of nociceptors, specialized sensory neurons that detect harmful stimuli like heat, pressure, or chemical changes. Once activated, these nociceptors send signals through **afferent nerve fibers** to the spinal cord, specifically to the **dorsal horn**. From here, pain signals are transmitted to the brain via several key pathways.

The most prominent pathway is the **spinothalamic tract**. In this pathway, pain signals are relayed from the dorsal horn of the spinal cord to the **thalamus**, a critical relay station in the brain. The thalamus processes and distributes these signals to various regions of the brain, including the **somatosensory cortex**, **limbic system**, and **prefrontal cortex**.

The **somatosensory cortex** is responsible for the localization and intensity of pain. When you stub your toe, the somatosensory cortex helps you identify the exact location and severity of the pain. This area of the brain creates a detailed map of your body, known as the **homunculus**, allowing for precise identification of where the pain originates.

The **limbic system**, which includes structures like the **amygdala** and **hippocampus**, processes the emotional aspects of pain. This is why pain is not just a physical sensation but also an emotional experience. The amygdala, in particular, is involved in the fear and anxiety that often accompany pain, while the hippocampus helps link pain to memories, which can influence future responses to similar stimuli.

The **prefrontal cortex** plays a role in the cognitive aspects of pain, such as evaluating its significance and deciding how to respond. This area is involved in attention and decision-making, helping you focus on the pain and take action, such as withdrawing from the source of harm or seeking medical help.

In addition to the spinothalamic tract, the **spinoreticular tract** carries pain signals to the **reticular formation**, which is involved in the autonomic responses to pain, like changes in heart rate or blood pressure. The **spinomesencephalic tract** transmits pain signals to the **periaqueductal gray (PAG)** in the midbrain, which plays a critical role in the modulation and suppression of pain.

Together, these pathways integrate the sensory, emotional, and cognitive components of pain, creating a complex experience that influences behavior and survival.

Modulation of Pain

Pain is not a static experience; it can be modulated by various factors, including psychological states, attention, and physiological mechanisms. The brain has the ability to amplify or diminish the perception of pain, a process known as **pain modulation**.

One of the primary systems involved in pain modulation is the **descending pain control system**, which originates in the brain and projects down to the spinal cord. Key regions involved in this system include the **periaqueductal gray (PAG)** in the midbrain, the **rostral ventromedial medulla (RVM)**, and the **dorsolateral pontine tegmentum**. These areas send inhibitory signals down the spinal cord to reduce the transmission of pain signals.

The PAG, in particular, is a crucial player in pain modulation. It receives input from various brain regions, including the prefrontal cortex and the amygdala, and uses this information to modulate pain based on the context. For example, in life-threatening situations, the PAG can trigger a powerful pain-suppressing response, allowing an individual to focus on survival rather than the pain. This is known as **stress-induced analgesia**.

Endogenous opioids, such as **endorphins**, **enkephalins**, and **dynorphins**, are chemicals produced by the body that play a significant role in pain modulation. These opioids bind to receptors in the brain and spinal cord, reducing the perception of pain. They are part of the body's natural pain control system and can be released during activities such as exercise, which is why physical activity can sometimes alleviate pain.

Attention and emotional state also influence pain perception. **Distraction** can reduce the perception of pain by shifting attention away from the pain stimulus. Conversely, **anxiety** and **catastrophizing** can amplify pain by increasing attention to it and anticipating worse outcomes. The prefrontal cortex plays a key role in this cognitive modulation of pain, helping to either focus on or divert attention from the pain.

Placebo effects are another fascinating aspect of pain modulation. When a person believes that a treatment will reduce their pain, the brain can release endogenous opioids and activate the descending pain control system, leading to real reductions in pain. This demonstrates the powerful connection between the mind and the perception of pain.

Overall, pain modulation is a dynamic process influenced by a complex interplay of neural, chemical, and psychological factors. Understanding these mechanisms is crucial for developing effective pain management strategies.

Pain Management

Pain management is the process of alleviating or reducing pain to improve a person's quality of life. Effective pain management requires a comprehensive approach that addresses both the physical and psychological aspects of pain. The choice of treatment often depends on the type, severity, and underlying cause of the pain.

Pharmacological treatments are a cornerstone of pain management. **Nonsteroidal anti-inflammatory drugs (NSAIDs)**, such as ibuprofen and aspirin, are commonly used to treat mild to moderate pain, particularly when inflammation is a contributing factor. These medications work by inhibiting enzymes involved in the production of **prostaglandins**, which are chemicals that promote inflammation and pain.

Opioids, such as morphine, oxycodone, and fentanyl, are potent pain relievers used for moderate to severe pain, particularly in acute or cancer-related pain. Opioids work by binding to specific receptors in the brain and spinal cord, reducing the perception of pain. However, they carry a risk of dependence, tolerance, and side effects, making them suitable only for certain types of pain and often for short-term use.

Adjuvant medications are drugs not primarily designed to treat pain but can be effective in certain pain conditions. **Antidepressants**, particularly **tricyclic antidepressants** and **serotonin-norepinephrine reuptake inhibitors (SNRIs)**, can help manage neuropathic pain by altering neurotransmitter levels in the brain and spinal cord. **Anticonvulsants**, such as gabapentin and pregabalin, are also used to treat neuropathic pain by stabilizing nerve activity.

Non-pharmacological treatments play a critical role in pain management, especially for chronic pain. **Physical therapy** helps to improve mobility, strengthen muscles, and reduce pain through targeted exercises and techniques. **Cognitive-behavioral therapy (CBT)** is another effective approach, addressing the psychological aspects of pain by changing the way a person thinks about and responds to their pain.

Interventional procedures, such as nerve blocks, epidural injections, and spinal cord stimulation, are options for patients with severe or refractory pain. These procedures aim to disrupt pain signals at various points in the nervous system, providing relief that may not be achievable through medication alone.

Lifestyle modifications are also essential in managing chronic pain. Regular exercise, a balanced diet, and adequate sleep can reduce pain levels and improve overall well-being. **Mind-body practices**, such as meditation, yoga, and relaxation techniques, can help reduce stress, enhance coping mechanisms, and lower the perception of pain.

Multidisciplinary pain management programs combine medical, physical, and psychological therapies to address all aspects of pain. These programs are particularly effective for chronic pain conditions, providing patients with a comprehensive set of tools to manage their pain and improve their quality of life.

Advances in pain management continue to evolve, with ongoing research into new medications, therapies, and interventions. Personalized pain management plans, tailored to the individual's specific needs and circumstances, offer the best outcomes for those suffering from acute or chronic pain.

CHAPTER 12: NEUROIMMUNOLOGY: THE BRAIN AND THE IMMUNE SYSTEM

The Blood-Brain Barrier

The blood-brain barrier (BBB) is a critical structure that protects the brain from harmful substances while allowing essential nutrients to pass through. This selective barrier is formed by tightly packed endothelial cells lining the blood vessels in the brain, along with supporting cells such as astrocytes and pericytes. The BBB is important in maintaining the brain's delicate environment, ensuring that it remains stable and free from toxins, pathogens, and fluctuations in blood-borne substances.

Structure and Function of the Blood-Brain Barrier

The endothelial cells of the BBB are unique compared to those in other parts of the body. They are connected by **tight junctions**, which are specialized protein complexes that seal the spaces between the cells. These tight junctions prevent most molecules from passing between the cells, forcing substances to pass through the cells themselves if they want to enter the brain. This selective permeability is what makes the BBB so effective at controlling what enters and leaves the brain's environment.

In addition to endothelial cells, **astrocytes** play a vital role in the function of the BBB. Astrocytes are star-shaped glial cells that surround the brain's blood vessels, providing support and signaling to the endothelial cells to maintain their tight junctions. Astrocytes also help regulate the passage of ions and nutrients, such as glucose and amino acids, from the blood into the brain. They are essential for maintaining the homeostasis of the brain's microenvironment.

Pericytes, another type of cell associated with the BBB, wrap around the endothelial cells and contribute to the barrier's structural integrity and stability. They are involved in the regulation of blood flow within the brain's microvasculature and play a role in repairing the barrier after injury.

Selective Permeability

The BBB is highly selective in what it allows to pass through. Small, lipophilic molecules, such as oxygen, carbon dioxide, and certain hormones, can diffuse across the BBB relatively easily. Water also crosses the barrier through specific channels known as **aquaporins**. However, larger molecules and hydrophilic substances require specific transport mechanisms to cross the BBB.

For example, glucose, the brain's primary energy source, is transported across the BBB by **glucose transporters** (GLUT1). Similarly, essential amino acids, which are necessary for protein synthesis and neurotransmitter production, are transported by **amino acid transporters**. These transporters ensure that the brain receives the nutrients it needs while preventing harmful substances from entering.

Protective Role

One of the most important functions of the BBB is to protect the brain from toxins, pathogens, and fluctuations in blood-borne substances that could disrupt neural function. For instance, many drugs, chemicals, and immune cells are unable to cross the BBB, which helps prevent them from causing damage to the brain. This protection is crucial because the brain is highly sensitive to changes in its environment, and even minor disruptions can have significant effects on neural activity.

However, the protective nature of the BBB also presents challenges, particularly in the treatment of neurological diseases. Many therapeutic drugs are unable to cross the BBB, making it difficult to deliver them to the brain in effective concentrations. Researchers are actively exploring ways to bypass or temporarily disrupt the BBB to allow for drug delivery without compromising the brain's safety.

BBB Dysfunction and Disease

When the BBB is compromised, it can lead to a range of neurological disorders. **BBB breakdown** is implicated in conditions such as multiple sclerosis, Alzheimer's disease, and stroke. In these cases, the barrier becomes more permeable, allowing harmful substances and immune cells to enter the brain, leading to inflammation, neural damage, and disease progression.

In multiple sclerosis, for example, the immune system attacks the myelin sheath of neurons, and this attack is facilitated by a compromised BBB that allows immune cells to infiltrate the central nervous system. In Alzheimer's disease, a weakened BBB may allow toxic proteins, such as beta-amyloid, to accumulate in the brain, contributing to neurodegeneration.

Understanding the BBB's function and how it can be disrupted provides insights into the development and progression of neurological diseases. It also highlights the importance of maintaining BBB integrity for overall brain health and the challenges that come with treating conditions that involve the central nervous system.

Neuroinflammation

Neuroinflammation is the brain's immune response to various stimuli, such as infection, injury, or disease. While inflammation in the brain is a natural defense mechanism designed to protect neural tissue, chronic or excessive neuroinflammation can lead to damage and contribute to the progression of neurological disorders.

The brain's immune response is primarily mediated by **microglia**, the resident immune cells of the central nervous system. Microglia constantly survey the brain for signs of damage or infection. When they detect a threat, they become activated, changing shape and releasing a range of signaling molecules, including **cytokines**, **chemokines**, and **reactive oxygen species**. These molecules help to coordinate the immune response by recruiting other immune cells to the site of injury and promoting the repair of damaged tissue.

Astrocytes, another type of glial cell, also play a role in neuroinflammation. In response to injury or disease, astrocytes can become reactive, a process known as **astrogliosis**. Reactive astrocytes help to limit the spread of inflammation by forming a protective barrier around the affected area. However, prolonged activation of astrocytes can lead to the formation of a **glial scar**, which can inhibit the regeneration of neurons and disrupt normal brain function.

While acute neuroinflammation is typically protective, chronic neuroinflammation can be harmful. Prolonged inflammation can lead to the release of neurotoxic substances that damage neurons and other cells in the brain. This chronic state of inflammation is implicated in several neurodegenerative diseases, including **Alzheimer's disease**, **Parkinson's disease**, and **multiple sclerosis**.

In Alzheimer's disease, for example, the accumulation of **beta-amyloid plaques** triggers a persistent inflammatory response. Microglia and astrocytes attempt to clear these plaques but, over time, become chronically activated, releasing substances that further damage neurons and exacerbate the disease.

In multiple sclerosis, the immune system attacks the myelin sheath that insulates nerve fibers, leading to neuroinflammation and demyelination. This results in the disruption of electrical signals in the brain and spinal cord, leading to the symptoms associated with the disease.

Understanding neuroinflammation is crucial for developing new treatments for neurological disorders. Therapeutic strategies that modulate the brain's immune response, reduce chronic inflammation, or promote resolution of inflammation are actively being researched as potential ways to protect the brain from neurodegenerative diseases and promote recovery after injury.

The Gut-Brain Axis

The gut-brain axis is the bidirectional communication network between the gastrointestinal (GI) tract and the brain. This complex system involves neural, hormonal, and immunological signaling pathways that allow the gut and brain to influence each other's function. Understanding the gut-brain axis has provided new insights into how the digestive system impacts mental health and vice versa.

One of the key components of the gut-brain axis is the **vagus nerve**, a major nerve that extends from the brainstem to the abdomen. The vagus nerve acts as a communication highway, transmitting signals between the gut and the brain. It's important in regulating digestive processes, such as the release of digestive enzymes and gut motility. Additionally, the vagus nerve helps to modulate the brain's response to stress and influences emotional and cognitive functions.

Another important aspect of the gut-brain axis is the **gut microbiota**, the trillions of microorganisms that reside in the GI tract. The gut microbiota plays a significant role in maintaining gut health, supporting digestion, and producing essential nutrients like vitamins and short-chain fatty acids. Recent research has shown that the gut microbiota also influences brain function and behavior. For example, certain gut bacteria produce neurotransmitters like **serotonin** and **gamma-aminobutyric acid (GABA)**, which are involved in regulating mood and anxiety.

Disruptions in the gut microbiota, known as **dysbiosis**, have been linked to a range of mental health conditions, including depression, anxiety, and irritable bowel syndrome (IBS). Dysbiosis can lead to increased intestinal permeability, often referred to as "leaky gut," allowing harmful substances to enter the bloodstream and potentially trigger inflammation in the brain. This inflammation can contribute to the development of mood disorders and cognitive decline.

The gut-brain axis also involves the **hypothalamic-pituitary-adrenal (HPA) axis**, a central stress response system. Stress can alter gut function by affecting the release of hormones and neurotransmitters, which in turn can influence gut motility, secretion, and permeability. Chronic stress is known to impact the composition of the gut microbiota, further linking mental health and gut health.

Research on the gut-brain axis has led to new approaches in treating mental health disorders, such as the use of **probiotics** and **prebiotics** to support a healthy gut microbiome. These interventions aim to restore balance in the gut microbiota, potentially reducing symptoms of anxiety, depression, and other related conditions.

The gut-brain axis highlights the importance of a holistic approach to health, where the well-being of the gut and brain are interconnected. Understanding this relationship opens up new possibilities for therapies that target both physical and mental health through the gut-brain connection.

Autoimmune Diseases of the Brain

Autoimmune diseases of the brain occur when the immune system mistakenly attacks the brain's own cells, leading to inflammation, damage, and a range of neurological symptoms. These diseases are often chronic and can be challenging to diagnose and treat due to their complexity and the overlap of symptoms with other neurological disorders.

One of the most well-known autoimmune diseases of the brain is **multiple sclerosis (MS)**. In MS, the immune system targets the myelin sheath, the protective covering that insulates nerve fibers in the central nervous system (CNS). This attack leads to demyelination, which disrupts the normal transmission of electrical impulses along the nerves. Over time, this damage can result in a wide range of symptoms, including muscle weakness, vision problems, difficulty with coordination and balance, and cognitive impairments. The course of MS can vary widely, with some individuals experiencing periods of relapse and remission, while others may have a more progressive form of the disease.

Neuromyelitis optica (NMO) is another autoimmune disease that primarily affects the optic nerves and spinal cord. Unlike MS, NMO is characterized by severe attacks that can lead to blindness and paralysis. The immune system in NMO specifically targets **aquaporin-4**, a water channel protein found in astrocytes. This results in inflammation and damage to the optic nerves and spinal cord, leading to the characteristic symptoms of the disease.

Autoimmune encephalitis is a group of conditions where the immune system attacks specific receptors or proteins in the brain, leading to inflammation of the brain tissue. One of the most studied forms is **anti-NMDA receptor encephalitis**, where antibodies target NMDA receptors, which are crucial for synaptic transmission and plasticity. This condition can cause a wide array of symptoms, including psychiatric disturbances (such as hallucinations and psychosis), seizures, memory loss, and movement disorders. Early recognition and treatment are crucial, as autoimmune encephalitis can be life-threatening if not promptly managed.

Guillain-Barré syndrome (GBS), though primarily affecting the peripheral nervous system, is another autoimmune disorder that can involve the brain, particularly in its severe forms, leading to **Miller Fisher syndrome** or other variants that affect cranial nerves and brainstem functions. GBS is often triggered by infections, where the immune system, in an attempt to fight the infection, mistakenly targets peripheral nerves, leading to muscle weakness and, in some cases, respiratory failure.

Autoimmune diseases of the brain require a multidisciplinary approach to diagnosis and treatment. Immunomodulatory therapies, such as corticosteroids, intravenous immunoglobulins (IVIG), and plasmapheresis, are commonly used to reduce inflammation and suppress the abnormal immune response. Early intervention is key to managing these conditions effectively and minimizing long-term neurological damage.

Neuroendocrine-Immune Interactions

The neuroendocrine-immune system represents the intricate and bidirectional communication network between the nervous system, endocrine system, and immune system. This interaction is essential for maintaining homeostasis and responding to internal and external stressors, including infections, injuries, and psychological stress.

One of the central components of this interaction is the **hypothalamic-pituitary-adrenal (HPA) axis**. The HPA axis is activated in response to stress, leading to the release of **corticotropin-releasing hormone (CRH)** from the hypothalamus. CRH stimulates the pituitary gland to secrete **adrenocorticotropic hormone (ACTH)**, which in turn triggers the adrenal glands to release **cortisol**, a key stress hormone. Cortisol has potent anti-inflammatory effects, suppressing the immune response and preventing excessive inflammation. However, chronic activation of the HPA axis, as seen in prolonged stress, can lead to immune dysregulation, making the body more susceptible to infections and autoimmune diseases.

The **sympathetic nervous system (SNS)** also plays a significant role in modulating immune function. During the fight-or-flight response, the SNS releases **norepinephrine**, which can directly influence immune cells, including lymphocytes, macrophages, and dendritic cells. Norepinephrine can enhance or suppress immune responses depending on the context, modulating the body's defense mechanisms against pathogens.

In addition to these direct pathways, the neuroendocrine-immune interactions involve the release of **cytokines** by immune cells, which can affect brain function. For example, during an infection, pro-inflammatory cytokines like **interleukin-1 (IL-1)**, **interleukin-6 (IL-6)**, and **tumor necrosis factor-alpha (TNF-α)** are released into the bloodstream. These cytokines can cross the blood-brain barrier and influence the brain's activity, leading to symptoms commonly associated with sickness behavior, such as fatigue, fever, and depression.

On the other hand, the nervous system can regulate immune function through neurotransmitters and neuropeptides. **Acetylcholine**, released by the parasympathetic nervous system, has anti-inflammatory effects by binding to the **alpha-7 nicotinic acetylcholine receptor** on immune cells, inhibiting the release of pro-inflammatory cytokines. This mechanism is known as the **cholinergic anti-inflammatory pathway**.

Hormones also play a critical role in neuroendocrine-immune interactions. **Estrogens** and **androgens** can modulate immune responses, which partly explains the differences in immune function and susceptibility to autoimmune diseases between men and women. For instance, estrogens generally enhance immune responses, which may contribute to the higher prevalence of autoimmune diseases in women, while androgens tend to have immunosuppressive effects.

Dysregulation of neuroendocrine-immune interactions is implicated in various disorders, including chronic inflammatory conditions, autoimmune diseases, and mental health disorders such as depression and anxiety. Understanding these interactions provides insights into the mechanisms underlying these conditions and offers potential targets for therapeutic intervention, aiming to restore balance across these interconnected systems.

CHAPTER 13: NEURODEGENERATIVE DISEASES

Alzheimer's Disease

Alzheimer's disease is a progressive neurodegenerative disorder that primarily affects memory, thinking, and behavior. It is the most common cause of dementia, accounting for 60-80% of cases. The disease gradually destroys brain cells, leading to a decline in cognitive abilities and, eventually, the loss of the ability to carry out basic daily activities.

Pathophysiology of Alzheimer's Disease

Alzheimer's disease is characterized by the accumulation of two abnormal protein aggregates in the brain: **beta-amyloid plaques** and **neurofibrillary tangles**. These proteins disrupt normal cellular function and lead to neuronal death.

Beta-amyloid plaques are sticky clumps of protein fragments that accumulate between neurons. These plaques form when a protein called amyloid precursor protein (APP) is broken down improperly, resulting in beta-amyloid fragments that are not cleared from the brain effectively. Over time, these fragments aggregate into plaques that disrupt communication between neurons and trigger inflammatory responses, further damaging brain tissue.

Neurofibrillary tangles are twisted fibers of a protein called **tau** found inside neurons. In healthy neurons, tau helps stabilize microtubules, which are essential for maintaining the cell's structure and transporting nutrients and other substances within the cell. In Alzheimer's disease, tau becomes abnormally phosphorylated, causing it to detach from microtubules and clump together into tangles. This disrupts the neuron's transport system, leading to cell death.

The **hippocampus**, a brain region critical for memory formation, is one of the first areas affected by Alzheimer's disease. As the disease progresses, it spreads to other regions of the brain, including the cerebral cortex, which is responsible for language, reasoning, and social behavior. This widespread neuronal loss leads to the characteristic symptoms of Alzheimer's, including memory loss, confusion, disorientation, and changes in personality.

Genetic and Environmental Factors

While the exact cause of Alzheimer's disease is not fully understood, it is believed to result from a combination of genetic, environmental, and lifestyle factors. The most well-known genetic risk factor is the presence of the **APOE ε4 allele**. Individuals with one or two copies of this allele have a higher risk of developing Alzheimer's, particularly at an earlier age. However, not everyone with this allele develops the disease, and people without it can still develop Alzheimer's, suggesting that other factors are also involved.

Environmental and lifestyle factors, such as education level, physical activity, and cardiovascular health, also play a role in the risk of developing Alzheimer's. There is evidence that a healthy diet, regular exercise, mental stimulation, and social engagement may help reduce the risk or delay the onset of the disease.

Symptoms and Diagnosis

The early symptoms of Alzheimer's disease often include subtle memory loss, particularly difficulty remembering recent events or conversations. As the disease progresses, symptoms become more pronounced, including difficulty with language, problem-solving, and completing familiar tasks. Later stages of the disease are marked by severe cognitive impairment, loss of communication abilities, and the need for full-time care.

Diagnosis of Alzheimer's disease is typically based on a combination of clinical assessments, cognitive tests, and neuroimaging. **Magnetic resonance imaging (MRI)** and **positron emission tomography (PET)** scans can reveal patterns of brain atrophy and detect the presence of amyloid plaques and tau tangles. However, a definitive diagnosis can only be confirmed through a post-mortem examination of brain tissue.

Current Treatments and Research

There is currently no cure for Alzheimer's disease, and treatment focuses on managing symptoms and improving quality of life. **Cholinesterase inhibitors**, such as donepezil and rivastigmine, are commonly prescribed to help maintain levels of acetylcholine, a neurotransmitter involved in memory and learning. **Memantine**, another medication, works by regulating the activity of glutamate, another neurotransmitter that can cause damage when overactive.

Research into Alzheimer's disease is ongoing, with efforts focused on understanding the underlying mechanisms of the disease and developing new treatments. Potential therapies under investigation include drugs that target beta-amyloid and tau proteins, as well as approaches aimed at reducing inflammation and protecting neurons from further damage.

Alzheimer's disease remains a major public health challenge, affecting millions of individuals and their families worldwide. Continued research and advances in understanding the disease will be vital in developing more effective treatments and, ultimately, finding a cure.

Parkinson's Disease

Parkinson's disease is a progressive neurodegenerative disorder that primarily affects movement. It results from the gradual loss of dopamine-producing neurons in a part of the brain called the **substantia nigra**, which is located in the midbrain. Dopamine is a neurotransmitter that plays a critical role in coordinating smooth and controlled muscle movements. As dopamine levels decrease, the brain's ability to regulate movement diminishes, leading to the characteristic symptoms of Parkinson's disease.

The primary motor symptoms of Parkinson's disease include **tremor, bradykinesia (slowness of movement), rigidity**, and **postural instability**. The tremor often starts in one hand and is most noticeable at rest. Bradykinesia makes simple tasks, like buttoning a shirt or writing, increasingly difficult as movements become slow and deliberate. Muscle stiffness, or rigidity, can cause discomfort and reduce the range of motion, while postural instability leads to balance problems and an increased risk of falls.

In addition to these motor symptoms, Parkinson's disease can also lead to non-motor symptoms, such as **cognitive impairment, depression, sleep disturbances**, and **autonomic dysfunction**. These non-motor symptoms can significantly impact the quality of life and often emerge as the disease progresses.

The exact cause of Parkinson's disease is not fully understood, but it is believed to result from a combination of genetic and environmental factors. Several genetic mutations have been associated with familial forms of Parkinson's, including mutations in the **LRRK2, PINK1, DJ-1**, and **SNCA** genes. Environmental factors, such as exposure to pesticides and heavy metals, have also been implicated in increasing the risk of developing the disease.

Lewy bodies, abnormal clumps of a protein called **alpha-synuclein**, are often found in the brains of people with Parkinson's disease. These aggregates are thought to contribute to the death of neurons in the substantia nigra, although their exact role in the disease process remains unclear.

Currently, there is no cure for Parkinson's disease, but several treatments are available to manage its symptoms. The most common treatment is **levodopa**, a precursor to dopamine that can cross the blood-brain barrier and be converted into dopamine in the brain. Levodopa is often combined with **carbidopa** to prevent its breakdown before reaching the brain. Other treatments include **dopamine agonists**, which mimic the effects of dopamine, and **MAO-B inhibitors**, which slow the breakdown of dopamine in the brain.

In some cases, **deep brain stimulation (DBS)**, a surgical procedure that involves implanting electrodes in specific brain regions, may be recommended to help control motor symptoms in patients who do not respond well to medication. DBS can significantly improve quality of life for some patients, although it is not suitable for everyone.

Research into Parkinson's disease continues, with a focus on understanding the underlying mechanisms of neurodegeneration, developing new treatments, and exploring potential neuroprotective strategies to slow or halt the progression of the disease.

Huntington's Disease

Huntington's disease is a hereditary neurodegenerative disorder that causes the progressive breakdown of nerve cells in the brain, leading to severe physical, cognitive, and psychiatric symptoms. The disease is caused by a mutation in the **HTT gene** on chromosome 4, which leads to the production of an abnormal form of the huntingtin protein. This abnormal protein accumulates in neurons, particularly in the **basal ganglia** and the **cortex**, leading to their dysfunction and eventual death.

Huntington's disease is characterized by a triad of symptoms: **motor dysfunction**, **cognitive decline**, and **psychiatric disturbances**. The most recognizable motor symptom is **chorea**, which involves involuntary, jerky movements of the limbs, face, and trunk. As the disease progresses, these movements become more pronounced and can interfere with daily activities, such as walking, eating, and speaking. In later stages, motor symptoms may evolve into **dystonia** (sustained muscle contractions) and **bradykinesia** (slowness of movement), making voluntary movements increasingly difficult.

Cognitive decline in Huntington's disease often begins with subtle changes in thinking and memory. Patients may experience difficulties with attention, planning, and decision-making, which can progress to more severe impairments resembling dementia. These cognitive changes can significantly affect an individual's ability to work, manage daily tasks, and maintain relationships.

Psychiatric symptoms are also a core feature of Huntington's disease. Depression, anxiety, irritability, and mood swings are common, and some patients may develop obsessive-compulsive behaviors or psychosis. These psychiatric symptoms can be as debilitating as the motor and cognitive symptoms and often contribute to the social isolation and reduced quality of life experienced by patients with Huntington's disease.

Huntington's disease is inherited in an **autosomal dominant** manner, meaning that a person only needs one copy of the mutated gene to develop the disease. Each child of an affected parent has a 50% chance of inheriting the mutation and developing the disease. The severity and age of onset are influenced by the number of CAG repeats in the HTT gene; a higher number of repeats is associated with an earlier onset and more severe progression.

Currently, there is no cure for Huntington's disease, and treatment focuses on managing symptoms and improving quality of life. Medications such as **tetrabenazine** and **antipsychotics** can help control chorea and psychiatric symptoms, while **antidepressants** and **mood stabilizers** may be used to treat depression and mood disorders. **Speech therapy**, **occupational therapy**, and **physical therapy** are also important components of care, helping patients maintain function and independence for as long as possible.

Research into Huntington's disease is ongoing, with efforts focused on understanding the disease mechanisms, developing new therapies, and exploring potential gene-editing technologies, such as **CRISPR**, to correct the genetic mutation at the root of the disease. While these advancements offer hope for future treatments, Huntington's disease remains a devastating condition with profound impacts on patients and their families.

Amyotrophic Lateral Sclerosis (ALS)

Amyotrophic Lateral Sclerosis (ALS), also known as Lou Gehrig's disease, is a progressive neurodegenerative disorder that primarily affects motor neurons— the nerve cells responsible for controlling voluntary muscles. The disease leads to the degeneration and eventual death of these motor neurons in the brain and spinal cord, resulting in a loss of muscle control and, ultimately, paralysis.

The early symptoms of ALS often include **muscle weakness**, **twitching (fasciculations)**, and **cramps**, which typically begin in one limb or a specific muscle group. As the disease progresses, the weakness spreads to other parts of the body, leading to difficulties with speaking (dysarthria), swallowing (dysphagia), and breathing. The disease affects both the **upper motor neurons** (which originate in the brain and send signals to the spinal cord) and **lower motor neurons** (which transmit signals from the spinal cord to the muscles), leading to a combination of spasticity (muscle stiffness) and muscle atrophy (wasting).

Sporadic ALS is the most common form of the disease, with no clear genetic link, while **familial ALS** accounts for about 5-10% of cases and is inherited in an autosomal dominant manner. Mutations in several genes, such as **SOD1**, **C9orf72**, and **TARDBP**, have been identified in familial ALS, but the exact mechanisms by which these mutations lead to motor neuron degeneration are not fully understood. It is believed that a combination of factors, including oxidative stress, mitochondrial dysfunction, and abnormal protein aggregation, contributes to the disease process.

As ALS progresses, patients may lose the ability to walk, speak, eat, and eventually breathe, requiring ventilatory support. However, the disease typically does not affect cognitive functions, sensory nerves, or bladder control, although some patients may develop a form of dementia known as **frontotemporal dementia (FTD)**, which affects personality, behavior, and executive function.

Currently, there is no cure for ALS, and treatment focuses on managing symptoms, slowing disease progression, and improving quality of life. **Riluzole** is one of the few FDA-approved drugs for ALS, and it works by reducing the release of glutamate, a neurotransmitter that can be toxic to neurons in excessive amounts. Another drug, **edaravone**, is thought to reduce oxidative stress in neurons and has been shown to slow the decline of physical function in some patients.

Supportive care, including physical therapy, occupational therapy, speech therapy, and nutritional support, is essential in managing the symptoms and maintaining the highest possible level of function and comfort. The average life expectancy after diagnosis is 3 to 5 years, although some individuals live longer, particularly with early intervention and comprehensive care.

Multiple Sclerosis

Multiple Sclerosis (MS) is a chronic autoimmune disease that affects the central nervous system (CNS), including the brain and spinal cord. In MS, the immune system mistakenly attacks the **myelin sheath**, the protective covering that surrounds nerve fibers, leading to inflammation, demyelination, and subsequent damage to the nerves themselves. This disruption in nerve signaling results in a wide range of neurological symptoms that vary depending on the location and extent of the damage.

MS is characterized by periods of **relapse** (acute worsening of symptoms) followed by **remission** (partial or complete recovery), although the disease course can vary. There are several types of MS, with the most common being **relapsing-remitting MS (RRMS)**, where patients experience distinct episodes of neurological dysfunction. Over time, RRMS can progress to **secondary progressive MS (SPMS)**, where symptoms worsen gradually without clear relapses or remissions. **Primary progressive MS (PPMS)** is less common and involves a steady progression of symptoms from the onset without relapses.

The symptoms of MS are highly variable and can include **fatigue, muscle weakness, spasticity, numbness or tingling, vision problems, balance and coordination difficulties, bladder and bowel dysfunction**, and **cognitive impairment**. These symptoms often fluctuate, with some individuals experiencing mild symptoms while others face significant disability.

The exact cause of MS is unknown, but it is believed to involve a combination of genetic predisposition and environmental factors, such as viral infections or low vitamin D levels. The immune system's attack on myelin is thought to be triggered by these factors, leading to the chronic inflammation and neurodegeneration seen in the disease.

Diagnosis of MS typically involves a combination of clinical evaluation, **magnetic resonance imaging (MRI)** to detect lesions in the CNS, **lumbar puncture** to analyze cerebrospinal fluid for immune markers, and **evoked potential tests** to assess the speed of nerve signal transmission.

There is no cure for MS, but several **disease-modifying therapies (DMTs)** are available to reduce the frequency and severity of relapses, slow disease progression, and limit the accumulation of new CNS lesions. These therapies include **interferon beta, glatiramer acetate, fingolimod, natalizumab**, and newer oral and infusion medications. In addition to DMTs, symptom management with physical therapy, medications for spasticity and pain, and lifestyle modifications are essential for improving quality of life.

Research into MS continues to focus on better understanding the disease mechanisms, developing more effective treatments, and finding potential cures. With early diagnosis and appropriate treatment, many individuals with MS can manage their symptoms and maintain a relatively active and fulfilling life.

Prion Diseases

Prion diseases, also known as transmissible spongiform encephalopathies (TSEs), are a group of rare, fatal neurodegenerative disorders caused by abnormally folded proteins known as prions. Unlike typical infectious agents like bacteria or viruses, prions are misfolded forms of a normal protein called **PrP (prion protein)**. When these misfolded prions come into contact with the normal version of the protein, they induce it to misfold as well, leading to a chain reaction that results in the accumulation of prions in the brain.

The accumulation of these abnormal proteins causes severe damage to brain tissue, leading to the characteristic spongy appearance under a microscope, hence the term "spongiform." As prions build up, they lead to the death of neurons, causing progressive neurological symptoms.

Prion diseases affect both humans and animals. In humans, the most common prion disease is **Creutzfeldt-Jakob disease (CJD)**, which can occur in several forms: sporadic (sCJD), inherited (familial CJD), or acquired through exposure to prions, such as through contaminated medical equipment or infected meat (variant CJD). Symptoms of CJD include rapidly progressive dementia, memory loss, personality changes, motor dysfunction, and, in later stages, severe cognitive decline and involuntary movements. The disease progresses rapidly, and most patients succumb within a year of symptom onset.

Another well-known prion disease is **kuru**, which was endemic among the Fore people of Papua New Guinea due to ritualistic cannibalism. The decline in these practices led to the near elimination of kuru. In animals, **bovine spongiform encephalopathy (BSE)**, commonly known as mad cow disease, is a prion disease that can be transmitted to humans through consumption of infected beef, leading to variant CJD.

There is currently no cure for prion diseases, and they are invariably fatal. Diagnosis is challenging and often involves ruling out other conditions through clinical evaluation, brain imaging, and analysis of cerebrospinal fluid for specific markers. A definitive diagnosis can only be made through brain tissue examination after death.

Research into prion diseases is ongoing, with a focus on understanding the mechanisms of prion propagation and identifying potential therapeutic strategies. Approaches being explored include targeting the misfolded prion protein for degradation, blocking its conversion from the normal form, and developing methods to diagnose prion diseases earlier.

Neuroprotective Strategies

Neuroprotective strategies aim to preserve neuronal function, prevent neuron death, and slow the progression of neurodegenerative diseases. These strategies are crucial in conditions where the brain's ability to protect and repair itself is overwhelmed, such as in Alzheimer's disease, Parkinson's disease, and amyotrophic lateral sclerosis (ALS).

One of the primary approaches to neuroprotection involves **antioxidants**, which help to reduce oxidative stress, a key factor in neuronal damage. Neurons are highly susceptible to oxidative stress because of their high metabolic rate and reliance on mitochondria for energy production. **Reactive oxygen species (ROS)**, which are byproducts of cellular metabolism, can damage DNA, proteins, and lipids, leading to cell death. Antioxidants like **vitamin E**, **vitamin C**, and **coenzyme Q10** can neutralize ROS, thereby protecting neurons from oxidative damage.

Anti-inflammatory agents also play a critical role in neuroprotection. Chronic inflammation in the brain, often driven by activated microglia and astrocytes, can contribute to neuronal injury and death. Drugs that reduce inflammation, such as **nonsteroidal anti-inflammatory drugs (NSAIDs)** or newer therapies targeting specific inflammatory pathways, have been studied for their potential to protect neurons in diseases like Alzheimer's.

Glutamate excitotoxicity is another major mechanism of neuronal damage, particularly in conditions like stroke and ALS. Glutamate is an essential neurotransmitter, but when present in excessive amounts, it overstimulates neurons, leading to cell death. Neuroprotective strategies aimed at reducing glutamate levels or blocking its receptors have shown promise in experimental models. **Memantine**, a drug used in Alzheimer's disease, works by blocking NMDA receptors, reducing the risk of excitotoxic damage.

Neurotrophic factors are proteins that support the growth, survival, and differentiation of neurons. Enhancing the levels or activity of these factors, such as **brain-derived neurotrophic factor (BDNF)** or **glial cell line-derived neurotrophic factor (GDNF)**, has been explored as a neuroprotective strategy. These factors can help repair damaged neurons, promote the formation of new connections, and protect against further injury.

Lifestyle interventions also play a role in neuroprotection. Regular physical exercise has been shown to enhance brain health by increasing blood flow, reducing inflammation, and promoting the release of neurotrophic factors. A diet rich in antioxidants, omega-3 fatty acids, and other neuroprotective compounds, such as those found in the Mediterranean diet, has been associated with a lower risk of neurodegenerative diseases.

Finally, ongoing research into gene therapy, stem cell therapy, and pharmacological agents aims to develop new and more effective neuroprotective strategies. While much of this research is still in the experimental stages, it holds the potential to significantly impact the treatment and management of neurodegenerative diseases, offering hope for preserving brain function and improving quality of life for affected individuals.

CHAPTER 14: NEUROPSYCHIATRIC DISORDERS

Schizophrenia

Schizophrenia is a complex and chronic neuropsychiatric disorder that profoundly affects how a person thinks, feels, and behaves. It is characterized by a range of symptoms, which can be broadly categorized into positive symptoms, negative symptoms, and cognitive impairments.

Positive Symptoms

Positive symptoms refer to experiences that are added to the person's normal functioning and include **hallucinations, delusions**, and **disorganized thinking**. Hallucinations are perceptions in the absence of external stimuli, with auditory hallucinations, such as hearing voices, being the most common. These voices often comment on the individual's behavior or give commands, which can be distressing and disruptive.

Delusions are strongly held false beliefs that are resistant to reason or contrary evidence. For example, a person with schizophrenia might believe they are being persecuted, that they have extraordinary abilities, or that they are being controlled by external forces. These delusions can significantly impact behavior and interactions with others.

Disorganized thinking manifests as incoherent or illogical speech, making it difficult for the person to communicate effectively. This might involve switching from one topic to another without a clear connection or responding with answers that are unrelated to the questions asked.

Negative Symptoms

Negative symptoms reflect a reduction or absence of normal behaviors and functions. These include **anhedonia** (a lack of pleasure in everyday activities), **avolition** (a lack of motivation to initiate and sustain activities), **alogia** (reduced speech output), and **flattened affect** (diminished emotional expression). Negative symptoms are particularly debilitating because they impair the person's ability to function in daily life, maintain relationships, and engage in social activities.

Cognitive Impairments

Cognitive impairments in schizophrenia affect various domains, including **attention, working memory, executive function**, and **processing speed**. These cognitive deficits contribute to the difficulties that individuals with schizophrenia face in maintaining employment, managing daily tasks, and engaging in social interactions. Unlike positive symptoms, which can fluctuate over time, cognitive impairments tend to be more stable and enduring, significantly impacting quality of life.

Neurobiological Factors

The exact cause of schizophrenia is not fully understood, but it is believed to involve a combination of genetic, environmental, and neurobiological factors. **Genetic predisposition** plays a significant role, with individuals who have a family history of schizophrenia being at higher risk of developing the disorder. However, genes alone are not sufficient to cause schizophrenia, and environmental factors such as prenatal stress, exposure to infections, or psychosocial stressors can contribute to its onset.

Neurobiologically, schizophrenia is associated with abnormalities in brain structure and function. One of the most well-known hypotheses is the **dopamine hypothesis**, which suggests that hyperactivity of dopamine transmission in certain brain regions, particularly the **mesolimbic pathway**, contributes to the positive symptoms of schizophrenia. This hypothesis is supported by the effectiveness of **antipsychotic medications**, which primarily work by blocking dopamine receptors, reducing the intensity of hallucinations and delusions.

In addition to dopamine, other neurotransmitters, including **glutamate** and **serotonin**, are believed to play a role in the disorder. Abnormalities in the **glutamatergic system**, particularly involving NMDA receptors, have been linked to both positive and negative symptoms. Brain imaging studies have also revealed structural changes in individuals with schizophrenia, such as enlarged **ventricles**, reduced **gray matter volume**, and altered connectivity in key brain networks, including the **prefrontal cortex** and **hippocampus**.

Treatment Approaches

Treatment for schizophrenia typically involves a combination of pharmacotherapy, psychotherapy, and psychosocial support. **Antipsychotic medications** are the cornerstone of treatment and are effective in reducing positive symptoms. However, they are less effective in addressing negative symptoms and cognitive impairments, which often require additional therapeutic approaches.

Cognitive-behavioral therapy (CBT) can help individuals manage symptoms by challenging delusional thinking and improving coping strategies. **Social skills training**, **vocational rehabilitation**, and **family therapy** are also important components of a comprehensive treatment plan, helping patients improve their functional abilities and social integration.

While schizophrenia is a chronic condition, early intervention and sustained treatment can significantly improve outcomes, allowing individuals to lead meaningful lives despite the challenges posed by the disorder. Continued research into the underlying mechanisms of schizophrenia holds promise for developing more effective treatments and ultimately improving the lives of those affected by this complex disorder.

Mood Disorders

Mood disorders encompass a group of mental health conditions characterized by significant disturbances in a person's emotional state. These disorders primarily affect mood, energy levels, and behavior, often leading to substantial impairments in daily functioning and quality of life. The two most common mood disorders are **major depressive disorder (MDD)** and **bipolar disorder**.

Major depressive disorder (MDD), also known as clinical depression, is marked by persistent feelings of sadness, hopelessness, and a lack of interest or pleasure in almost all activities. People with MDD may also experience changes in appetite, sleep disturbances (either insomnia or hypersomnia), fatigue, difficulty concentrating, and thoughts of death or suicide. The exact cause of MDD is not fully understood, but it is believed to result from a combination of genetic, biological, environmental, and psychological factors.

Neurobiological factors play a significant role in MDD. Dysregulation of neurotransmitters such as **serotonin, norepinephrine**, and **dopamine** has been implicated in the pathophysiology of depression. Brain imaging studies have shown alterations in the structure and function of specific brain regions, including the **prefrontal cortex, amygdala**, and **hippocampus**, which are involved in mood regulation and cognitive processes.

Bipolar disorder is characterized by extreme mood swings that include emotional highs (mania or hypomania) and lows (depression). During a manic episode, individuals may feel euphoric, full of energy, and unusually irritable. They may engage in risky behaviors, experience racing thoughts, and have a reduced need for sleep. Conversely, depressive episodes in bipolar disorder are similar to those seen in MDD, with feelings of deep sadness, hopelessness, and a lack of energy.

Bipolar disorder is divided into **Bipolar I** and **Bipolar II**. Bipolar I involves episodes of full-blown mania, while Bipolar II is characterized by hypomania (a milder form of mania) and depressive episodes. The exact causes of bipolar disorder are not well understood, but genetic factors are known to play a strong role, with family history being a significant risk factor.

Treatment for mood disorders typically involves a combination of **pharmacotherapy** and **psychotherapy**. Antidepressants, such as **selective serotonin reuptake inhibitors (SSRIs)**, are commonly used to treat MDD, while **mood stabilizers** and **antipsychotic medications** are often prescribed for bipolar disorder. **Cognitive-behavioral therapy (CBT)** and **interpersonal therapy (IPT)** are effective psychotherapeutic approaches for managing mood disorders, helping patients develop coping strategies, challenge negative thought patterns, and improve their emotional regulation.

While mood disorders are chronic conditions, early intervention and consistent treatment can lead to significant improvements in symptoms and overall quality of life.

Anxiety Disorders

Anxiety disorders are a group of mental health conditions characterized by excessive, persistent, and often irrational fear or worry. These disorders are among the most common psychiatric conditions, affecting millions of people worldwide. Anxiety disorders can significantly impair a person's ability to function in daily life and are often accompanied by physical symptoms such as a racing heart, sweating, trembling, and gastrointestinal issues.

There are several types of anxiety disorders, with the most common being **generalized anxiety disorder (GAD)**, **panic disorder**, **social anxiety disorder**, and **specific phobias**.

Generalized anxiety disorder (GAD) is characterized by chronic, excessive worry about various aspects of life, such as work, health, or relationships. This worry is often difficult to control and is accompanied by physical symptoms like restlessness, muscle tension, and difficulty concentrating. People with GAD may find it challenging to relax and often expect the worst outcomes in everyday situations.

Panic disorder involves recurrent, unexpected panic attacks—sudden episodes of intense fear that reach a peak within minutes. Symptoms of a panic attack can include chest pain, shortness of breath, dizziness, and a fear of losing control or dying. The fear of having another panic attack often leads to significant behavioral changes, such as avoiding certain places or situations.

Social anxiety disorder, also known as social phobia, is characterized by an intense fear of social situations where the individual might be judged, embarrassed, or scrutinized by others. This fear can lead to avoidance of social interactions, severely impacting the person's ability to form relationships, work, or participate in social activities.

Specific phobias are intense, irrational fears of specific objects or situations, such as heights, animals, or flying. The fear experienced is often out of proportion to the actual danger posed by the object or situation and can lead to avoidance behaviors that disrupt daily life.

The underlying causes of anxiety disorders are thought to involve a combination of genetic, environmental, and neurobiological factors. Dysregulation of neurotransmitters such as **serotonin, norepinephrine,** and **gamma-aminobutyric acid (GABA)** is associated with anxiety. Additionally, brain regions like the **amygdala** and **prefrontal cortex**, which are involved in processing fear and regulating emotions, show altered activity in individuals with anxiety disorders.

Treatment for anxiety disorders typically includes a combination of **medication** and **psychotherapy**. **Selective serotonin reuptake inhibitors (SSRIs)** and **benzodiazepines** are commonly prescribed to help manage symptoms. **Cognitive-behavioral therapy (CBT)** is particularly effective in treating anxiety disorders by helping individuals challenge and change maladaptive thought patterns and behaviors, gradually reducing their anxiety.

With appropriate treatment, many individuals with anxiety disorders can manage their symptoms and lead fulfilling lives. Early intervention and a comprehensive treatment plan are important for improving outcomes and enhancing quality of life.

Autism Spectrum Disorder

Autism Spectrum Disorder (ASD) is a neurodevelopmental disorder characterized by challenges in social communication and interaction, as well as restricted, repetitive patterns of behavior, interests, or activities. The term "spectrum" reflects the wide range of symptoms and severity seen in individuals with ASD, from highly functioning individuals to those with significant cognitive impairments and communication difficulties.

One of the core features of ASD is **difficulty with social communication**. Individuals with ASD may struggle to understand and respond to social cues, such as facial expressions, body language, and tone of voice. This can lead to challenges in forming relationships, making eye contact, and engaging in typical back-and-forth conversations. Some individuals with ASD may also have difficulty understanding and expressing emotions, which can further complicate social interactions.

Restricted and repetitive behaviors are another hallmark of ASD. These behaviors can include repetitive movements, such as hand-flapping or rocking, as well as intense, focused interests in specific topics, objects, or activities. For example, a child with ASD might become deeply interested in trains, memorizing detailed information about different models and schedules. Changes in routine or environment can be particularly distressing for individuals with ASD, leading to anxiety and behavioral outbursts.

The exact cause of ASD is not fully understood, but it is believed to result from a combination of **genetic** and **environmental factors**. Numerous genes have been associated with ASD, and these genetic variations likely interact with environmental influences, such as prenatal exposure to toxins or infections, to increase the risk of developing the disorder.

Brain imaging studies have shown differences in the structure and function of certain brain regions in individuals with ASD. For example, abnormalities in the **amygdala**, which is involved in processing emotions, and the **prefrontal cortex**, which is important for social cognition, have been observed. These differences may contribute to the social and communication challenges seen in ASD.

Early diagnosis and intervention are crucial for improving outcomes in individuals with ASD. Behavioral therapies, such as **Applied Behavior Analysis (ABA)**, are commonly used to teach social, communication, and adaptive skills. Speech and occupational therapies can also play a significant role in helping individuals with ASD develop the skills needed for daily life. In some cases, medications may be prescribed to manage symptoms such as anxiety, aggression, or hyperactivity.

While ASD is a lifelong condition, many individuals with ASD can lead fulfilling lives with appropriate support and intervention. Increasing awareness, acceptance, and understanding of ASD is essential for creating inclusive environments that allow individuals with ASD to thrive.

Personality Disorders

Personality disorders are a group of mental health conditions characterized by enduring patterns of behavior, cognition, and inner experience that deviate significantly from cultural expectations. These patterns are pervasive, inflexible, and lead to distress or impairment in personal, social, or occupational functioning. Personality disorders typically manifest in adolescence or early adulthood and are categorized into three clusters based on similar characteristics.

Cluster A personality disorders are characterized by odd or eccentric behaviors. This cluster includes **paranoid personality disorder, schizoid personality disorder**, and **schizotypal personality disorder**. Individuals with these disorders may exhibit distrust, social detachment, or unusual beliefs and behaviors. For example, someone with schizotypal personality disorder might have strange, magical thinking or believe they have special powers.

Cluster B personality disorders involve dramatic, emotional, or erratic behavior. This cluster includes **borderline personality disorder (BPD), narcissistic personality disorder, histrionic personality disorder**, and **antisocial**

personality disorder. People with BPD, for instance, often experience intense emotional instability, fear of abandonment, and difficulties in maintaining relationships. In contrast, individuals with narcissistic personality disorder may display grandiosity, a need for admiration, and a lack of empathy.

Cluster C personality disorders are characterized by anxious or fearful behaviors. This cluster includes **avoidant personality disorder, dependent personality disorder**, and **obsessive-compulsive personality disorder (OCPD)**. Individuals with avoidant personality disorder often experience extreme social inhibition, feelings of inadequacy, and sensitivity to negative evaluation. On the other hand, those with OCPD may be preoccupied with orderliness, perfectionism, and control, often at the expense of flexibility and efficiency.

The causes of personality disorders are believed to be multifactorial, involving a combination of **genetic, biological**, and **environmental factors**. Childhood experiences, such as trauma, neglect, or inconsistent parenting, are thought to contribute to the development of these disorders. Genetic predispositions may also play a role, as personality traits can be inherited.

Treatment for personality disorders often involves **psychotherapy** as the primary approach. **Cognitive-behavioral therapy (CBT), dialectical behavior therapy (DBT)**, and **psychodynamic therapy** are commonly used to help individuals understand their thoughts and behaviors, develop healthier coping mechanisms, and improve interpersonal relationships. Medications may be prescribed to manage specific symptoms, such as mood instability, anxiety, or depression, although they are not typically the primary treatment for personality disorders.

Personality disorders can be challenging to treat due to the ingrained nature of the behaviors and thought patterns. However, with long-term therapy and support, individuals can make significant improvements in their functioning and quality of life. Understanding and recognizing the impact of personality disorders is crucial for providing appropriate care and support to those affected.

Addiction and the Brain

Addiction is a chronic and relapsing disorder characterized by compulsive drug-seeking behavior, loss of control over use, and the emergence of negative emotional states when access to the drug is prevented. Understanding addiction requires a comprehensive look at the neurobiological processes that occur in the brain, particularly how repeated exposure to addictive substances alters brain circuits involved in reward, motivation, and self-control.

The Brain's Reward System

At the core of addiction lies the **brain's reward system**, which is primarily mediated by the neurotransmitter **dopamine**. The reward system involves several key brain structures, including the **ventral tegmental area (VTA), nucleus accumbens**, and **prefrontal cortex**. When a person engages in pleasurable activities, such as eating or socializing, dopamine is released from the VTA and sent to the nucleus accumbens, creating a sense of pleasure and reinforcing the behavior.

Addictive substances, such as drugs and alcohol, hijack this reward system by causing a much more intense and prolonged release of dopamine than natural rewards. For example, drugs like cocaine directly increase dopamine levels by blocking its reuptake, leading to an accumulation of dopamine in the synapse. This flood of dopamine creates a powerful sense of euphoria, strongly reinforcing the drug-taking behavior.

Changes in the Brain with Chronic Use

With repeated exposure to addictive substances, the brain undergoes significant changes. One of the key changes is **tolerance**, where the brain becomes less responsive to the drug's effects, requiring higher doses to achieve the same level of euphoria. This is partly due to a reduction in the number of dopamine receptors in the nucleus accumbens, as the brain attempts to maintain balance in response to the overstimulation caused by the drug.

Another major change is the **sensitization** of the brain's reward pathways to the cues associated with drug use. Environmental cues, such as places, people, or paraphernalia associated with drug use, become highly potent triggers for craving and relapse. These cues activate the **amygdala** and **hippocampus**, which are involved in emotional memory and stress responses, leading to intense urges to seek the drug.

The **prefrontal cortex**, which is involved in decision-making, impulse control, and executive function, also becomes compromised with chronic drug use. This area of the brain loses its ability to regulate the reward system effectively, leading to a loss of control over drug-seeking behaviors despite the negative consequences. This dysfunction in the prefrontal cortex is a key reason why addiction is often described as a disease of choice, where the individual's ability to choose to stop using the drug is severely impaired.

The Role of Stress and Negative Affect

As addiction progresses, the brain's reward system becomes less responsive to the drug, and the individual begins to experience withdrawal symptoms when the drug is not available. These symptoms, which can include anxiety, irritability, and dysphoria, are mediated by the **extended amygdala**, a brain region involved in stress and negative emotions. The individual may begin to use the drug not to achieve pleasure but to avoid the discomfort of withdrawal, leading to a vicious cycle of dependence.

The **hypothalamic-pituitary-adrenal (HPA) axis**, the body's central stress response system, also becomes dysregulated in addiction. Chronic drug use can lead to an overactive stress response, which further exacerbates negative emotional states and drives compulsive drug-seeking behavior as a means of self-medication.

Recovery and Neuroplasticity

Despite the profound changes in the brain caused by addiction, recovery is possible. The brain retains a remarkable ability to change and adapt, a property known as **neuroplasticity**. With sustained abstinence, the brain's reward system and prefrontal cortex can gradually recover, although this process can take time and may require a combination of therapeutic approaches.

Treatment for addiction often involves **behavioral therapies** such as **cognitive-behavioral therapy (CBT)**, which helps individuals identify and change maladaptive thought patterns and behaviors. **Medications** can also be effective in managing withdrawal symptoms and reducing cravings, particularly for substances like opioids and alcohol.

Understanding addiction as a brain disorder underscores the importance of a comprehensive, evidence-based approach to treatment that addresses both the biological and psychological aspects of the disease. With the right support, individuals can overcome addiction and rebuild their lives.

CHAPTER 15: THERAPEUTIC APPROACHES IN NEUROSCIENCE

Pharmacotherapy

Pharmacotherapy is a cornerstone of treatment in neuroscience, involving the use of medications to manage, alleviate, or cure neurological and psychiatric conditions. The brain's complex network of neurotransmitters, receptors, and neural circuits provides multiple targets for pharmacological intervention, making it possible to address a wide range of disorders.

Neurotransmitter Modulation

One of the primary mechanisms of pharmacotherapy in neuroscience is the modulation of neurotransmitter systems. Neurotransmitters are chemicals that transmit signals between neurons, playing crucial roles in regulating mood, cognition, and behavior. Many psychiatric and neurological disorders are associated with imbalances in these chemicals.

For example, **selective serotonin reuptake inhibitors (SSRIs)** are widely used in the treatment of depression and anxiety disorders. SSRIs work by blocking the reuptake of serotonin, a neurotransmitter associated with mood regulation, into the presynaptic neuron. This increases the availability of serotonin in the synaptic cleft, enhancing its action on postsynaptic receptors and alleviating symptoms of depression.

Similarly, **dopamine agonists** are used in the treatment of Parkinson's disease, a disorder characterized by the degeneration of dopamine-producing neurons in the substantia nigra. Medications like **levodopa** increase dopamine levels in the brain, helping to restore motor function and reduce symptoms such as tremor and rigidity.

Targeting Receptors and Ion Channels

Pharmacotherapy also targets specific receptors and ion channels to modulate neural activity. **Benzodiazepines**, for instance, enhance the action of **gamma-aminobutyric acid (GABA)**, the primary inhibitory neurotransmitter in the brain. By binding to GABA-A receptors, benzodiazepines increase the influx of chloride ions into neurons, hyperpolarizing the cell membrane and reducing neuronal excitability. This mechanism makes benzodiazepines effective in treating conditions like anxiety, insomnia, and epilepsy.

Antipsychotic medications, used in the treatment of schizophrenia and bipolar disorder, primarily target **dopamine D2 receptors**. These drugs work by blocking dopamine receptors in the brain's mesolimbic pathway, reducing the hyperactivity of dopamine that is thought to contribute to the positive symptoms of schizophrenia, such as hallucinations and delusions.

Calcium channel blockers are another class of drugs that modulate neural activity by inhibiting the flow of calcium ions through voltage-gated calcium channels. These medications are used in the management of conditions like epilepsy and migraine, where they help stabilize neuronal membranes and prevent excessive firing of neurons.

Disease-Modifying Therapies

In addition to symptom management, pharmacotherapy in neuroscience increasingly focuses on **disease-modifying therapies** that aim to alter the underlying disease process. In multiple sclerosis (MS), for example, **disease-modifying drugs (DMDs)** like **interferon beta** and **fingolimod** work by modulating the immune system to reduce the frequency of relapses and slow the progression of disability. These therapies do not cure MS but can significantly improve the long-term outlook for patients by protecting the central nervous system from ongoing damage.

Similarly, in Alzheimer's disease, research is ongoing to develop drugs that target the **amyloid plaques** and **tau tangles** that are characteristic of the disease. While current treatments like **cholinesterase inhibitors** and **memantine** provide symptomatic relief, future therapies may focus on preventing or clearing these pathological features to slow or halt disease progression.

Personalized Medicine

Advances in pharmacogenomics are paving the way for **personalized medicine** in neuroscience. By understanding the genetic factors that influence an individual's response to medications, clinicians can tailor treatment plans to maximize efficacy and minimize adverse effects. For example, genetic testing can identify patients who are more likely to respond to certain antidepressants or who are at increased risk for side effects from antipsychotic medications.

Personalized medicine holds great promise for improving outcomes in neurological and psychiatric disorders, as it allows for more precise and targeted interventions based on an individual's unique genetic makeup and neurobiological profile.

Challenges and Future Directions

Despite the advances in pharmacotherapy, challenges remain, including the need for better treatments for conditions that are currently difficult to manage, such as treatment-resistant depression and neurodegenerative diseases. Research continues to explore new targets and mechanisms, with the goal of developing more effective and safer therapies.

As our understanding of the brain's complexities deepens, pharmacotherapy will likely continue to evolve, offering new hope for individuals affected by neurological and psychiatric disorders.

Neurosurgery

Neurosurgery is a specialized field of medicine that involves the surgical treatment of disorders affecting the brain, spinal cord, and peripheral nerves. It is a crucial therapeutic approach in neuroscience, used to address a wide range of conditions, from brain tumors and traumatic injuries to epilepsy and movement disorders.

One of the most common neurosurgical procedures is the removal of **brain tumors**. Tumors can be either benign (non-cancerous) or malignant (cancerous), and their location and size can significantly impact neurological function. Neurosurgeons use advanced imaging techniques, such as **magnetic resonance imaging (MRI)** and **computed tomography (CT)** scans, to precisely locate the tumor. During surgery, tools like **microscopes** and **neuronavigation systems** help the surgeon remove the tumor while minimizing damage to surrounding healthy brain tissue. In some cases, neurosurgeons may use **awake brain surgery** to ensure critical areas involved in speech or movement are preserved.

Neurosurgery also plays a vital role in the treatment of **epilepsy**, particularly in patients who do not respond to medication. **Resective surgery** involves removing the part of the brain where seizures originate, such as the temporal lobe in cases of **temporal lobe epilepsy**. Another approach, **vagus nerve stimulation (VNS)**, involves implanting a device that sends electrical impulses to the vagus nerve, reducing the frequency and severity of seizures.

For patients with **movement disorders** like **Parkinson's disease**, neurosurgery offers a highly effective treatment option called **deep brain stimulation (DBS)**. DBS involves implanting electrodes in specific brain regions, such as the **subthalamic nucleus** or **globus pallidus**, to regulate abnormal electrical activity. The electrodes are connected to a pacemaker-like device implanted in the chest, which delivers continuous electrical stimulation to help control symptoms like tremors, rigidity, and bradykinesia.

Neurosurgery is also critical in the management of **traumatic brain injuries (TBI)** and **spinal cord injuries**. In cases of severe TBI, surgery may be necessary to relieve pressure on the brain caused by swelling, bleeding, or fractures. **Craniotomy**, where a portion of the skull is temporarily removed, allows the surgeon to access the brain and perform the necessary repairs. Similarly, in spinal cord injuries, neurosurgery can stabilize the spine, remove bone fragments, or decompress the spinal cord to prevent further damage.

Despite its effectiveness, neurosurgery carries risks, including infection, bleeding, and neurological deficits. Advances in surgical techniques, imaging, and post-operative care have significantly improved outcomes and reduced complications. However, neurosurgery is often a last resort, typically pursued when other treatments have failed or when the condition poses an immediate threat to the patient's life or quality of life.

Psychotherapy and Behavioral Interventions

Psychotherapy and behavioral interventions are essential components of treatment in neuroscience, particularly for mental health disorders such as depression, anxiety, and post-traumatic stress disorder (PTSD). These therapeutic approaches focus on modifying dysfunctional thoughts, behaviors, and emotions, helping individuals develop healthier coping mechanisms and improve their overall well-being.

One of the most widely used forms of psychotherapy is **cognitive-behavioral therapy (CBT)**. CBT is based on the idea that negative thought patterns and beliefs contribute to emotional distress and maladaptive behaviors. By identifying and challenging these thoughts, patients can learn to replace them with more balanced and realistic perspectives. CBT is effective for a variety of conditions, including depression, anxiety disorders, and obsessive-compulsive disorder (OCD). It typically involves structured sessions where the therapist and patient work collaboratively to set goals, practice new skills, and monitor progress.

Dialectical behavior therapy (DBT) is a specialized form of CBT developed for individuals with **borderline personality disorder (BPD)** and those who struggle with emotional regulation. DBT combines cognitive-behavioral techniques with concepts from mindfulness and acceptance. It focuses on teaching patients skills in four key areas: mindfulness, distress tolerance, emotion regulation, and interpersonal effectiveness. DBT has been shown to reduce self-harming behaviors, improve emotional stability, and enhance relationships.

Exposure therapy is another effective behavioral intervention, particularly for anxiety disorders such as **phobias**, **PTSD**, and **social anxiety disorder**. In exposure therapy, patients gradually confront the situations or objects that trigger their anxiety in a controlled and safe environment. The repeated exposure helps desensitize them to the fear, reducing the anxiety over time. Techniques like **systematic desensitization**, where exposure is paired with relaxation exercises, and **prolonged exposure**, where patients repeatedly recount traumatic memories, are commonly used.

Behavioral activation (BA) is a therapy aimed at treating depression by encouraging individuals to engage in positive and rewarding activities. Depression often leads to withdrawal and inactivity, which can worsen the symptoms. BA helps patients identify and schedule activities that align with their values and interests, thereby increasing positive experiences and improving mood. This approach emphasizes the role of behavior in influencing emotions, making it a practical and action-oriented treatment.

Mindfulness-based interventions are increasingly popular in the treatment of stress, anxiety, and depression. These approaches, such as **Mindfulness-Based Stress Reduction (MBSR)** and **Mindfulness-Based Cognitive Therapy (MBCT)**, involve practices like meditation, breathing exercises, and body awareness. Mindfulness teaches patients to observe their thoughts and feelings without judgment, fostering a greater sense of acceptance and reducing the impact of negative emotions.

Psychotherapy and behavioral interventions are often used in combination with pharmacotherapy, providing a comprehensive approach to treatment. These therapies not only address the symptoms of mental health disorders but also equip individuals with the tools they need to manage their conditions in the long term. With the guidance of a trained therapist, patients can make significant progress toward recovery and improved quality of life.

Neuromodulation Techniques

Neuromodulation techniques involve the use of electrical, magnetic, or chemical stimulation to alter neural activity in the brain and nervous system. These techniques are increasingly used to treat a variety of neurological and psychiatric conditions, including chronic pain, depression, epilepsy, and Parkinson's disease. By modulating specific neural circuits, these interventions can alleviate symptoms, enhance function, and improve quality of life for patients.

One of the most well-known neuromodulation techniques is **deep brain stimulation (DBS)**. DBS involves surgically implanting electrodes into specific brain regions, such as the **subthalamic nucleus** or **globus pallidus**, which are implicated in movement disorders like Parkinson's disease. These electrodes deliver continuous electrical pulses that help regulate abnormal neural activity. DBS has been particularly effective in reducing tremors, rigidity, and other motor symptoms in Parkinson's patients, allowing for better movement control and reduced medication needs.

Transcranial magnetic stimulation (TMS) is another non-invasive neuromodulation technique used primarily to treat depression, especially in cases where patients have not responded to antidepressant medications. TMS uses magnetic fields to stimulate nerve cells in the brain, particularly in the **prefrontal cortex**, which is involved in mood regulation. During a TMS session, an electromagnetic coil is placed against the scalp, delivering brief magnetic pulses that pass through the skull and induce electrical currents in the brain. These currents modulate neural activity, which can lead to improvements in mood and cognitive function over time.

Vagus nerve stimulation (VNS) is a neuromodulation technique that involves sending electrical impulses to the vagus nerve, which runs from the brainstem to the abdomen. VNS is used to treat epilepsy and depression. For epilepsy, VNS helps reduce the frequency and severity of seizures by influencing neural circuits in the brainstem that control electrical activity. In depression, VNS is thought to modulate mood by affecting neurotransmitter systems and brain regions involved in emotional regulation.

Transcranial direct current stimulation (tDCS) is another non-invasive technique that applies a low electrical current to the scalp through electrodes. This current modulates neuronal excitability, either enhancing or inhibiting activity in targeted brain areas. tDCS is being explored as a treatment for conditions such as depression, anxiety, and chronic pain. It is relatively simple to administer and can be used in both clinical and research settings.

Spinal cord stimulation (SCS) is primarily used to manage chronic pain conditions, such as neuropathic pain and failed back surgery syndrome. In SCS, electrodes are implanted in the epidural space of the spinal cord, where they deliver electrical pulses that interfere with pain signals traveling to the brain. Patients can control the intensity of the stimulation using an external device, providing them with relief from pain.

Neuromodulation techniques represent a growing field in neuroscience, offering new avenues for treating conditions that are difficult to manage with traditional therapies. As research advances, these techniques are likely to become even more refined and targeted, providing greater benefits to patients.

Cognitive Rehabilitation

Cognitive rehabilitation is a therapeutic approach designed to help individuals who have experienced brain injury or neurological conditions regain or improve cognitive functions such as memory, attention, problem-solving, and executive function. This form of therapy is crucial for patients recovering from traumatic brain injury (TBI), stroke, multiple sclerosis, and other conditions that impair cognitive abilities.

Cognitive rehabilitation involves a variety of strategies and techniques tailored to the specific needs and goals of the patient. One key aspect of cognitive rehabilitation is **restorative training**, which focuses on improving specific cognitive functions through repetitive practice and exercises. For example, a patient with memory impairments might engage in memory drills, use mnemonics, or practice recalling information in increasingly complex tasks. The goal is to strengthen the neural networks involved in these cognitive processes, promoting recovery of function.

Another important component is **compensatory strategies**, which help patients work around their cognitive deficits by developing new ways of completing tasks. For instance, a patient with attention deficits might learn to break tasks into smaller, more manageable steps or use external aids such as planners, alarms, or checklists to stay organized and focused. These strategies do not necessarily restore lost functions but help patients adapt to their new cognitive limitations and maintain independence.

Metacognitive training is also a crucial element of cognitive rehabilitation. This approach encourages patients to become more aware of their cognitive strengths and weaknesses, enabling them to self-monitor and adjust their behavior accordingly. For example, a patient with executive function difficulties might learn to set specific goals, monitor their progress, and make adjustments as needed to achieve those goals. This self-awareness can lead to better decision-making and problem-solving in daily life.

In addition to individual exercises and strategies, cognitive rehabilitation often involves **computer-based training programs** that provide structured cognitive exercises tailored to the patient's needs. These programs can target specific cognitive domains, such as attention, memory, or processing speed, and allow for personalized, intensive training.

Cognitive rehabilitation is typically delivered by a multidisciplinary team, including neuropsychologists, occupational therapists, speech-language pathologists, and other specialists. This team approach ensures that all aspects of the patient's cognitive and functional abilities are addressed, leading to more comprehensive and effective rehabilitation.

The effectiveness of cognitive rehabilitation depends on various factors, including the severity of the cognitive deficits, the patient's motivation, and the timing of the intervention. Early intervention is generally associated with better outcomes, as the brain's plasticity is greater in the initial stages of recovery.

Cognitive rehabilitation is an essential part of the recovery process for individuals with cognitive impairments, helping them regain function, improve quality of life, and return to daily activities. By combining restorative training, compensatory strategies, and metacognitive techniques, cognitive rehabilitation provides a comprehensive approach to managing the challenges of cognitive deficits.

APPENDIX

Terms and Definitions

1. **Neuron**: A nerve cell that is the basic building block of the nervous system.
2. **Synapse**: The junction between two neurons where communication occurs.
3. **Neurotransmitter**: A chemical substance that transmits signals across a synapse.
4. **Action Potential**: A rapid rise and fall in voltage across a cell membrane, resulting in a nerve impulse.
5. **Myelin Sheath**: A fatty layer that surrounds axons, increasing the speed of nerve transmission.
6. **Dendrite**: Branch-like structures of a neuron that receive signals from other neurons.
7. **Axon**: The long, slender projection of a neuron that conducts electrical impulses away from the cell body.
8. **Glial Cell**: Non-neuronal cells that provide support and protection for neurons.
9. **Cerebrum**: The largest part of the brain, responsible for voluntary activities, sensory perception, and cognition.
10. **Cerebellum**: A brain structure responsible for coordination and balance.
11. **Brainstem**: The structure connecting the brain to the spinal cord, responsible for automatic survival functions.
12. **Cortex**: The outer layer of the cerebrum, involved in higher brain functions like thought and action.
13. **Hippocampus**: A brain structure involved in memory formation.
14. **Amygdala**: A brain structure involved in emotion, particularly fear and pleasure.
15. **Thalamus**: A relay station in the brain for sensory information.
16. **Hypothalamus**: A brain region involved in homeostasis, regulating hunger, thirst, and temperature.
17. **Basal Ganglia**: A group of structures involved in movement control.
18. **Spinal Cord**: The bundle of nerves that transmits messages between the brain and the body.
19. **Peripheral Nervous System (PNS)**: The part of the nervous system outside the brain and spinal cord.
20. **Central Nervous System (CNS)**: The brain and spinal cord, responsible for processing information.
21. **Neuroplasticity**: The brain's ability to reorganize itself by forming new neural connections.
22. **Neurogenesis**: The process of generating new neurons in the brain.

23. **Synaptic Plasticity**: The ability of synapses to strengthen or weaken over time.
24. **Long-Term Potentiation (LTP)**: A long-lasting increase in synaptic strength following high-frequency stimulation.
25. **Long-Term Depression (LTD)**: A long-lasting decrease in synaptic strength.
26. **Blood-Brain Barrier**: A selective barrier that protects the brain from harmful substances in the blood.
27. **Cerebrospinal Fluid (CSF)**: A clear fluid found in the brain and spinal cord that cushions and protects the CNS.
28. **Neurotransmission**: The process of communication between neurons.
29. **Ion Channel**: A protein that allows ions to pass through the cell membrane, critical for neuron function.
30. **Resting Membrane Potential**: The electrical charge difference across the neuron's membrane when it is not active.
31. **Refractory Period**: The period following an action potential during which a neuron is unable to fire again.
32. **Neurotransmitter Receptor**: A protein on the cell surface that binds to neurotransmitters and initiates a cellular response.
33. **Excitatory Post-Synaptic Potential (EPSP)**: A synaptic potential that makes a neuron more likely to fire an action potential.
34. **Inhibitory Post-Synaptic Potential (IPSP)**: A synaptic potential that makes a neuron less likely to fire an action potential.
35. **Acetylcholine (ACh)**: A neurotransmitter involved in muscle movement and memory.
36. **Dopamine**: A neurotransmitter involved in reward, motivation, and motor control.
37. **Serotonin**: A neurotransmitter involved in mood, sleep, and appetite.
38. **GABA (Gamma-Aminobutyric Acid)**: The primary inhibitory neurotransmitter in the brain.
39. **Glutamate**: The primary excitatory neurotransmitter in the brain.
40. **Norepinephrine**: A neurotransmitter involved in arousal and the fight-or-flight response.
41. **Endorphins**: Neurotransmitters involved in pain relief and feelings of pleasure.
42. **Neurotrophins**: Proteins that support the growth, survival, and differentiation of neurons.
43. **Oligodendrocyte**: A type of glial cell that produces myelin in the CNS.
44. **Astrocyte**: A type of glial cell involved in blood-brain barrier maintenance and nutrient support for neurons.
45. **Microglia**: Immune cells of the CNS that remove waste and protect against pathogens.
46. **Synaptogenesis**: The formation of new synapses between neurons.
47. **Neural Circuit**: A network of interconnected neurons that process specific types of information.
48. **Cortical Column**: A vertical organization of neurons in the cortex, believed to be the basic processing unit.

49. **Sensory Neuron**: A neuron that transmits sensory information from the body to the CNS.
50. **Motor Neuron**: A neuron that transmits signals from the CNS to muscles or glands.
51. **Interneuron**: A neuron that connects sensory and motor neurons within the CNS.
52. **Reflex Arc**: A simple neural circuit that mediates a reflex action.
53. **Homunculus**: A distorted representation of the human body in the brain, reflecting sensory or motor control.
54. **Limbic System**: A group of brain structures involved in emotion, memory, and motivation.
55. **Corpus Callosum**: A bundle of nerve fibers that connects the left and right hemispheres of the brain.
56. **Broca's Area**: A region in the frontal lobe involved in speech production.
57. **Wernicke's Area**: A region in the temporal lobe involved in language comprehension.
58. **Neuroimaging**: Techniques that visualize brain activity, such as MRI and fMRI.
59. **Electrophysiology**: The study of the electrical properties of biological cells and tissues, including neurons.
60. **EEG (Electroencephalography)**: A technique for recording electrical activity of the brain via electrodes placed on the scalp.
61. **PET Scan (Positron Emission Tomography)**: An imaging technique that measures metabolic activity in the brain.
62. **fMRI (Functional Magnetic Resonance Imaging)**: An imaging technique that measures brain activity by detecting changes associated with blood flow.
63. **Neuroethics**: The study of ethical issues related to neuroscience research and its applications.
64. **Cognitive Neuroscience**: The study of the neural mechanisms underlying cognition and behavior.
65. **Behavioral Neuroscience**: The study of how the brain affects behavior.
66. **Neuroeconomics**: An interdisciplinary field that studies decision-making by combining neuroscience, economics, and psychology.
67. **Neuropharmacology**: The study of how drugs affect the nervous system.
68. **Neuromodulation**: The physiological process by which a given neuron uses one or more chemicals to regulate diverse populations of neurons.
69. **Deep Brain Stimulation (DBS)**: A surgical procedure that involves implanting electrodes in the brain to treat neurological disorders.
70. **Optogenetics**: A technique that uses light to control neurons that have been genetically modified to express light-sensitive ion channels.
71. **Neurodegeneration**: The progressive loss of structure or function of neurons, including their death.
72. **Amyloid Plaques**: Aggregates of proteins that are associated with Alzheimer's disease.
73. **Neurofibrillary Tangles**: Twisted fibers of a protein called tau, found within neurons in Alzheimer's disease.

74. **Excitotoxicity**: The pathological process by which neurons are damaged and killed by excessive stimulation by neurotransmitters such as glutamate.
75. **Neurogenetics**: The study of the role of genetics in the development and function of the nervous system.
76. **Epigenetics**: The study of changes in gene expression that do not involve changes to the underlying DNA sequence, affecting how cells read genes.
77. **Pruning**: The process of eliminating excess neurons and synapses during development to improve the efficiency of neural circuits.
78. **Neurotransmitter Transporter**: A protein that moves neurotransmitters across cell membranes, either into or out of cells.
79. **Synaptic Vesicle**: A small sac that stores neurotransmitters in the presynaptic neuron and releases them into the synapse.
80. **Neural Network**: A group of interconnected neurons whose activity is coordinated to produce a specific function or behavior.

AFTERWORD

We hope that as you close this book, you're feeling a mix of accomplishment, wonder, and curiosity about the amazing organ that is your brain.

Step by step, we've explored its structures, functions, and mysteries. From the microscopic dance of neurotransmitters to the grand symphony of cognition and emotion, neuroscience provides insights into the very essence of what makes us human.

As you reflect on what you've learned, take a moment to marvel at the fact that the very organ you've been reading about is the one that has allowed you to understand, process, and remember this information. Your brain has been busy forming new neural connections, strengthening synapses, and perhaps even generating a few new neurons as you've engaged with these concepts. In a way, your brain has been reshaping itself through the very act of learning about itself – how meta is that?

Neuroscience is a rapidly evolving field, with new discoveries being made all the time. The knowledge you've gained from this book provides you with a solid foundation to continue exploring. Whether you're considering a career in neuroscience, looking to apply this knowledge in your current field, or simply satisfying your curiosity, there's always more to learn.

We encourage you to stay curious. Ask questions. Challenge assumptions. The brain is still full of mysteries.

Understanding the brain isn't just about scientific knowledge – it's about understanding ourselves and each other a little better. The insights you've gained into how our brains process information, generate emotions, and shape our behaviors can help foster empathy and improve your interactions with others.

As you move forward, we hope you'll look at the world around you with fresh eyes. Perhaps you'll notice how your brain processes sensory information as you walk down the street, or you'll think about memory consolidation as you drift off to sleep. Maybe you'll have a new perspective on the challenges faced by those with neurological or psychiatric disorders.

Studying neuroscience comes with the realization that the brain is not just a fascinating organ, but a gateway to better understanding ourselves and the world around us.

Made in the USA
Monee, IL
01 December 2024

71680878R00075